Principles of

Engineering Geology

(Revised Edition)

Principles of
Engineering Geology

(Revised Edition)

K.V.G.K. Gokhale
Formerly Professor & Head
Dept of Civil Engg., I.I.T., Kanpur

BSP BS Publications
A unit of **BSP Books Pvt., Ltd.**

4-4-309/316, Giriraj Lane, Sultan Bazar,
Hyderabad - 500 095
Phone : 040 - 23445605, 23445688

Published by :

BSP BS Publications
A unit of **BSP Books Pvt., Ltd.**

4-4-309/316, Giriraj Lane, Sultan Bazar,
Hyderabad - 500 095
Phone : 040 - 23445605, 23445688
e-mail : info@bspbooks.net

ISBN : 978-93-52300-65-5 (HB)

To

Avantika & Akshita

Preface

Engineering Geology deals with applications of geology in Civil Engineering practice. The term is used in this context only although Geology is applied in several other engineering disciplines such as Mining, Metallurgical, Chemical and Materials Engineering fields. Engineering Geology is a part of the Civil Engineering curriculum (a compulsory course for one or two-semester duration) in all the Civil Engineering Departments in India and abroad.

Geologists work hand-in-hand with civil engineers in site selection and ground treatment for the foundations of dams and for reservoirs, tunnel supports, highway, airport locations, etc. Geophysics, Rock Mechanics and Geotechnical Engineering have significantly contributed to site evaluation and engineering geological classifications of rock masses and weathered material. Besides the foundation aspects, geological knowledge is very essential in several branches of Civil Engineering. In the area of Water Resources Engineering, projects spanning from groundwater exploration to watershed development and groundwater pollution require sound geological understanding of the region under consideration. Geological knowledge is vital in tackling problems concerning natural hazards such as earthquakes and landslides. The causative factors can be identified and the disaster-management handled effectively with proper knowledge of Engineering Geology. The study of reservoir-induced seismicity has become necessary in India ever since the Koyna earthquake of 1967, leading to a general awareness of the side effects in constructing large dams in unsound geological terrains, especially in the active Himalayan valleys.

Availability of remotely-sensed satellite data has improved the quality and use of mapping techniques in engineering geological investigations. For example, time-based (temporal) satellite digital data facilitate delineation of areas of coastal erosion and monitoring of the advance of the coastal erosion front. Geographic Information Systems (GIS) as a tool for database management has been of immense help in Engineering Geology. Thematic maps can be prepared for each of the parameters involved in any of the problems. Need-specific maps, e.g., engineering geological, earthquake or landslide vulnerability, hazard zone and groundwater potential zone maps, are prepared using GIS. Similarly, Global Positioning System (GPS) is gaining importance as a tool for precise measurements of crustal movements in earthquake-tsunami affected regions and deflections of dams due to seasonal fluctuations of impounded water.

This book is designed for the use of Civil Engineering students. It is an outcome of author's teaching for about four decades at I.I.T., Kanpur and elsewhere. At the undergraduate stage, the students are exposed to this subject for the first time and hence adequate basic geological information has been provided while focusing on the engineering applications. Concepts have been presented in the text for creative thinking and appreciation by the students. The entire spectrum of Civil Engineering applications has been dealt with, in different chapters through case studies, mostly of Indian projects. An effort has been made to stress the relevance and to create awareness of the applied aspects, in various aspects of civil engineering practice. Background material that cannot fit into the main body of the text, but is important for understanding certain basics, has been provided in Annexures I and II. Review questions on each chapter provided in Annexure III at the end of the book will help students in assessing their understanding of the subject matter. Being an outdoor subject, it was felt essential to illustrate the field features through several photographs.

The author gratefully acknowledges generous and spontaneous help from several individuals and organizations for the preparation of this book. Prof. N.Barton and Prof Z.T.Bieniawski have been kind enough in permitting the use of their classification rating tables and graphs. The author appreciates the prompt and encouraging responses received from Prof. Pradeep Talwani, Prof.John McArthur, Prof. M.R.Madhav, Prof. F.M.N. Kulkarni, Prof.D.Chandrasekharam, Dr.J.Krishna Murthy, Dr Sailesh Nayak, Prof.V.N.Bajpai and Prof.A.M.Bhola for permitting the author to reproduce figures and graphs from their work. Photographs included in this book are from several individuals and organizations as cited underneath each of the figure. The author is thankful to each one of them for their help.

Portions of the manuscript have been reviewed by Prof M.R. Madhav, Mr. S.G. Sukumar, Prof. K.L.Narahari Rao and Prof.C.V.R.Murthy. The author appreciates their timely suggestions. Discussions held with Prof A.K.Keshari , Prof. P.Dayaratnam and Prof.S.Surya Rao have been very helpful . The author records his deep appreciation of continuous support and encouragement received from his wife Sita who not only has meticulously gone through different versions of the manuscript but has offered suggestions for its improvement. She has been a continuous source of inspiration all these years. Timely support received from Nagendra, Ramachandra, Lakshmi and Sharada in the preparation of the manuscript is acknowledged.

"Minor changes have been incorporated in this Revised Edition. The author appreciates the suggestions received from the readers in this direction."

Date: 25-11-2009 **- Author**

Contents

Chapter 2

Engineering Properties and Classification of Rock Masses

Chapter 3

Weathering and Soil Formation

Chapter 4

Geomorphic Processes and Landforms

Chapter 5

Crustal Deformation and Geological Structures

Chapter 6

Site Investigations and Ground Improvement

Chapter 7

Hydrogeology

Chapter 8

Engineering Geology in Dam Construction

Chapter 9

Engineering Geology in Tunneling Practice

Chapter 10

Geological Hazards

1

Minerals and Rocks

Minerals, rocks and soils constitute earth materials. These materials influence the processes active on the surface and in the subsurface of our planet. They play a vital role in the site evaluation and operations in Civil Engineering practice. Whether it is in tunnelling, hydroelectric projects, groundwater development, foundation treatments or in the assessment of slope stability, a basic understanding of the earth materials is essential. Thus the study of these materials forms the first step.

Minerals

A mineral is a naturally occurring inorganic compound with a definite chemical composition, physical properties and crystalline structure. To be considered as a mineral, all these attributes are necessary. Minerals are the building blocks for the rocks. One rock is distinguished from another essentially on the basis of its mineralogical composition. The different criteria for classification of various rock types will be dealt later.

Minerals are broadly grouped into (a) the rock-forming minerals and (b) ore-forming minerals. The former constitute a rock while the latter form the composition of an ore. In Civil Engineering practice, it is important to have knowledge of the important rock-forming types. The ore-forming minerals are to be understood in detail by the Mining, Metallurgical and Mineral Engineering professionals.

All the minerals are grouped into 8 classes :

1. Native elements (like gold, silver, copper, sulphur and carbon)
2. Sulphides
3. Oxides and hydroxides
4. Halides
5. Carbonates, nitrates and borates
6. Sulphates, chromates, molybdates and tungstates
7. Phosphates, arsenates and vanadates
8. Silicates

Among these, the minerals of interest to Civil Engineers are essentially the oxides, hydroxides, carbonates and silicates.

In a rock, among the minerals present, the ones, which are dominant and characterize the rock, are termed as the essential (primary) minerals. The remaining present in the rock are known as accessory minerals. The accessory minerals normally are present in relatively small proportions and are composed of the minor components.

Essential Rock-Forming Minerals

Important among this category are :

Silicates

- Quartz (silica)
- Feldspars (Na-, K- and Ca- aluminate silicates)
- Amphiboles (Na-, Ca-, Mg-, Fe- Al silicates)
- Pyroxenes (Mg-, Fe- and Ca- silicates)
- Micas (K-, Mg-, Fe- Al silicates)
- Garnets (Fe, Mg, Mn, Ca and Al silicates)
- Olivine (Mg and Fe silicate)
- Clay minerals (K, Fe, Mg and Al silicates)

Carbonates

- Calcite (Ca carbonate)
- Dolomite (Ca-Mg carbonate)

Minerals with Fe, Mg and Mn in their composition are dark-coloured. The dark-coloured ones are thus the ferro-magnesian minerals.

Silicates

Silicate minerals form the bulk (around 95%) of the Earth's crust. Of these silicate minerals, quartz and feldspar are the most common ones in the crust. With increasing depth (for instance in the mantle region), the minerals are of Fe, Mg silicates such as the pyroxenes and olivines.

In a silicate structure, Si-O tetrahedra are the units. Silicon has tetrahedral coordination with oxygen. This is dictated by the ratio of the radius of the cation (silicon) to the ratio of the anion (oxygen). This ratio is commonly termed as the radius ratio. In a closed packing of silicon with oxygen, only four oxygen ions can be in linkage with the centrally situated cation (silicon) forming a Si-O tetrahedron. For close packing, it is necessary that, while all the anions (oxygen) not only are in contact with each other, but also should individually be in contact with the central cation (silicon). In such a situation, only four of the oxygen ions can be in linkage with one silicon ion. If the cation involved is not silicon but Al, for its close-packing with oxygen or hydroxyl, $(OH)^-$, it is in octahedral coordination with six of these anions O^- or OH^- surrounding each Al^{3+} ion. In silicate minerals with other cations such as Al, K, Na, Ca, Mg and Fe, these silicon-oxygen tetrahedra occur in specific linkages with other cation-hydroxyl octahedra.

Silicate minerals are classified on the basis of the linkage of Si-O tetrahedra.

The different groups are :

			Si : O
1.	Nesosilicates	Independent tetrahedral groups (ex. forsterite : $Mg_2\ SiO_4$)	1 : 4
2.	Sorosilicates	Double tetrahedral structures (ex. hemimorphite: Zn-silicate)	2 : 7
3.	Inosilicates	Single chain structure (ex. pyroxene : $Mg\ SiO_3$)	1 : 3
		Double chain structure {ex . amphibole: $Mg_7\ (Si_4O_{11})_2(OH)_2$}	4 :11
4.	Phyllosilicates	Sheet structure (ex. kaolinite: $Al_4\ Si_4O_{10}\ (OH)_8$)	2 : 5
5.	Cyclosilicates	Closed rings (ex. beryl: $Be_3\ Al_2\ Si_6\ O_{18}$)	1 : 3
6.	Tektosilicates	Continuous framework of tetrahedra (ex. quartz : SiO_2)	1 : 2

The bonding between the silica tetrahedra and the other cation linkages impart the characteristics to the individual silicate minerals. For example, in the structure of quartz, the Si-O tetrahedra being in continuous framework are in strong bonding and consequently the mineral quartz has no cleavage. On the other hand, in the clay mineral structure (also in the micas), the Si-O sheet alternates with a sheet of Al- (OH) octahedral sheet and due to weak bonding between the different unit layers of the mineral, distinct basal cleavage exists. This is clearly seen in the case of mica, which can be split along the cleavage (parting). Minerals with chain structure are generally fibrous or prismatic.

Mineralogical Phase Rule

Although several hundreds of minerals exist in nature, maximum number possible in a rock is dictated by the components and the degrees of freedom in the parent material. As we know, components are the smallest number of independent chemical entities, which completely define the composition of the system. These can be elements or compounds since some of the elements exist together as compounds in the parent melt.

Minerals are the individual phases in the system. They exist as homogeneous within themselves and are different from each other. To find out the maximum number of phases, the number of components and degrees of freedom operative in the geological process of the mineral formation are to be understood and substituted in the phase rule expression.

Phase rule

In a system under equilibrium conditions

$$P + F = C + 2$$

P = number of phases
C = number of components
F = degrees of freedom

In a geological system, the degrees of freedom are :

- bulk composition
- temperature
- pressure

For a melt of a particular composition, the pressure and temperature are the two degrees of freedom

Hence, $P = C$ (for $F = 2$)

This expression is termed as the mineralogical phase rule.

Thus the maximum number of minerals (phases) will be equal to the number of components (in the composition of the melt).

For example, if we consider pure silica (SiO_2), at any specific P and T conditions, two phases only can coexist (like quartz and tridymite).

In an igneous melt, the possible components are O, Si, Al, K, Na, Ca, Fe and Mg. These are eight in number. Hence in general, the maximum number of minerals possible can be eight. However, the number will be much less since O exists in combination with others such as SiO_2.

Igneous Rocks

Igneous rocks form through cooling and crystallization of molten rock material. If this molten material is below the Earth's surface, it is called magma. If it comes out above the surface, it is known as lava.

Nature of Magma

The molten rock material is semi-solid in nature (like porridge) and consists of liquid, gas and earlier-formed crystals. The volatiles are dominantly water vapour and carbon-di-oxide constituting around 15% of the material by weight. The main elements are oxygen, silicon, aluminum, calcium, sodium, potassium, iron and magnesium.

The volatiles in the magma influence its fluidity, mobility and the melting point. Its viscosity is largely controlled by silicon and water. Magmas with more volatile material erupt more violently. Magma is mobile and moves through the rocks of the crust and is capable of penetrating or intruding into them. As it moves to upper levels, it gradually cools and solidifies into a rock with its constituent minerals crystallizing during this process.

Magmas can be classified into two types :

Silicic magma : Composed mainly of silica (around 65% or more) with a temperature below 800 °C . Because of the large presence of silica and relatively low temperature, it is thick and viscous. There is greater resistance to flow due to the Si-O tetrahedral linkages.

Basaltic magma : It has silica content less than 50% and temperature relatively higher than silicic magmas (more than about 900 °C). It is more fluidy and mobile. When it comes onto the surface, it spreads out. Most popular example in India is Deccan Trap lava flows, which occupy a substantial area in Maharashtra and Madhya Pradesh.

Sequence of Crystallization of Minerals in Magma

During cooling of the molten rock material, as the temperature falls down, minerals of basic composition crystallize first controlled by the respective melting points. Minerals rich in silica subsequently crystallize. At the end of the crystallization period, the excess silica that remains forms quartz. So quartz is the last mineral to crystallize. Thus, in the deeper levels within the Earth, the material is extremely basic (ultrabasic) in its composition. As the depth decreases, the basic rocks form followed by granites and similar silicic rocks in the upper parts of the crust.

This sequence has first been proposed by Bowen and hence known as the Bowen's Reaction Series. In this sequence, depending upon the discrete formation of minerals or a gradual change in composition from one to another, the discontinuous and the continuous reaction series are proposed. The series can be illustrated as follows :

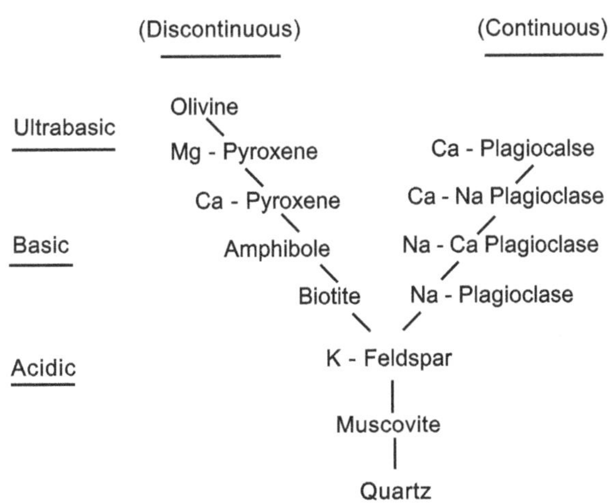

In the continuous reaction series, calcium-rich plagioclase (anorthite) appears first during cooling. With the gradual decrease in the temperature of the melt, its composition gradually changes into a more alkali-rich type. Between the pure Ca-rich plagioclase (anorthite) and alkali-rich plagioclase (albite), several plagioclase types with intermittent compositions are there (forming a solid-solution series). In the final stages, albite (alkali plagioclase) appears as a mineral.

From the above series, it can be seen that the ultrabasic rocks crystallize first followed by the basic and ultimately the acidic ones. Thus a granite (acidic rock) forming towards the end phase of crystallization is essentially composed of quartz, K-feldspar (orthoclase) and muscovite mica.

The temperature ranges for crystallization of the minerals from the melt are given in Table 1.1.

Table 1.1 Crystallization temperatures of important rock-forming minerals

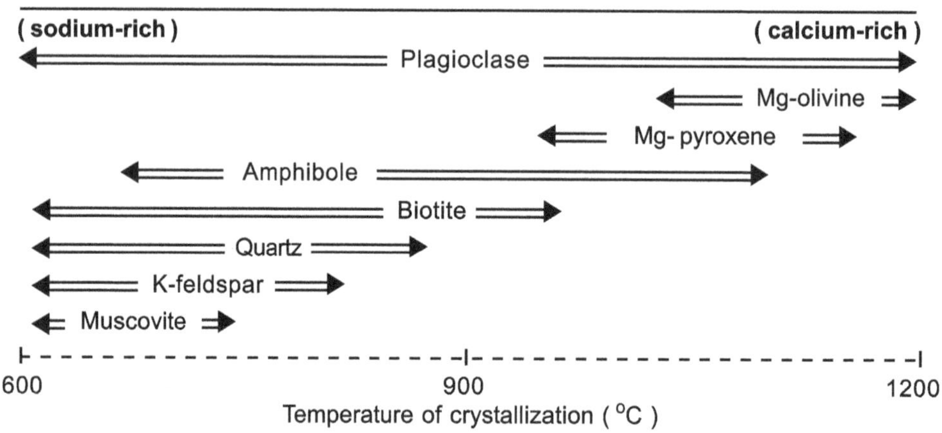

Classification of Igneous Rocks

Based on the silica content in its composition, an igneous rock is classified as :

- acidic
- intermediate
- basic
- ultrabasic

Acid igneous rocks are very rich in silica and poor in the ferromagnesian minerals. Quartz, alkali feldspar and mica are the common constituents. Presence of quartz indicates excess silica remaining after all the other chemical ingredients have combined with requisite proportions of silica to form minerals like feldspar and mica. Besides granites, pegmatites also fall in this category. Pegmatites are the last ones to form during magmatic crystallization conforming to the last stage in the Bowen's reaction series.

In the intermediate igneous rocks, free silica in the form of quartz is either less or absent. Alkali feldspar (the K-variety) is the main ingredient. Syenite and diorite pertain to this category.

The basic igneous rocks are composed essentially of the dark ferromagnesian minerals. Quartz is absent or if at all present it is in minute quantities. Plagioclase feldspar is significant in its proportion. Presence of plagioclase in the rock is indicative of relatively reduced silica content as compared to the acid igneous variety. Gabbro, dolerite and basalt are typical examples of this category.

The ultrabasic rocks are very deficient in silica. Quartz is invariably absent. The rocks are rich in Mg and the dominant ferromagnesian minerals contain magnesium. Peridotite and dunite are typical examples. Dunite essentially is monomineralic (composed of olivine as the dominant mineral).

Table 1.2 Generalized classification of igneous rocks

	ACIDIC	INTERMEDIATE		BASIC	ULTRABSIC
Coarse - grained	Granite	Syenite	Diorite	Gabbro	Peridotite Dunite
	Quartz porphyry	Syenite porphyry	Diorite porphyry	Dolerite	—
Fine-grained	Rhyolite	Trachyte	Andesite	Basalt	—
	AMORPHOUS			GLASS	—
Silica (%)	>65	55-65		45-65	< 45

Presence of all these four types, since their formation is temperature-pressure controlled, is depth-related. From the surface, as the depth increases, the acid igneous rocks are followed by the intermediate, basic and finally the ultrabasic types at great depths in the lower mantle. Earth is thus a differentiated planet with lighter rocks in the crust and the heavier ones at depths. The Earth's density thus increases from the surface to the core. The core, centrally located, is of very heavy material similar to nickel- iron metals in its composition (Annexure - II).

Texture and Mode of Occurrence of Igneous Rocks

Texture refers to the grain size, shape and mutual arrangement (fabric) of the grains. The igneous rocks formed at great depths are known as plutonic variety. Since cooling takes place over considerable period of time, the crystallization of minerals is slow and the grain growth continues unhampered. Thus the plutonic igneous rocks are coarse-grained. Compared to this situation, the volcanic igneous rocks that form on cooling and consolidation of lava (molten rock material above the surface) have very fine-grained texture. The temperature falls rapidly during cooling and consequently there is not enough time for crystallization to take place. The material is rapidly chilled. In fact, the solidification is so rapid that the volatile constituents escape forcing their way through channels during such rapid solidification. This results in the vesicular structure in the volcanic rocks. Any volcanic rock, for example basalt, is characterized by the presence of vesicles or interconnected pores. In some cases, these vesicular openings may be filled with minerals deposited from solutions at a later stage. These secondary minerals (like zeolites in the Deccan Trap basalts) in the vesicles are thus not genetically related to the volcanic rock in which they are present. The structure of the rock with infilled vesicles is termed as amygdoloidal structure.

The vesicular openings only host such minerals. Also, due to rapid cooling, tension cracks develop. These cracks are polygonal in shape and vertically extend downwards (Fig. 1.1). These planar openings are called columnar joints since on weathering, the material within these polygons stands out as vertical columns. Columnar joints are common in basalts. The rock in such a case is said to have columnar structure.

Fig. 1.1 Columnar joints in basalt, Mt. Baker, U.S.A (Photo : Nagendra Kolluru).

If the grains are in the same size range, the texture is referred to as equigranular. Otherwise it is termed as inequigranular. If the larger mineral grains have inclusions of the smaller ones, the texture of the rock is designated as poikilitic texture. In the case of dolerites, the texture is ophitic with laths of plagioclase enclosed by grains of pyroxene.

If large grains of minerals are surrounded by the fine-grained matrix, the texture of the rock is termed as porphyritic texture.

Mode of Occurrence of Intrusive Igneous Rocks

Igneous intrusions occur in different sizes and forms depending on the manner of intrusion. Dykes and sills are the common forms. If the intrusion is parallel to the layering in the host rock, it is called as a sill. On the other hand, if the intrusive is present cutting across the trend of the host rock, it is known as a dyke (Fig. 1.2).

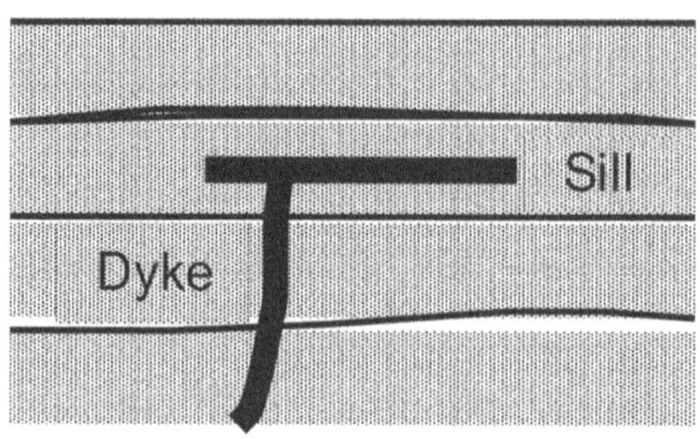

Fig. 1.2 Dyke and sill.

If the intrusion takes place forcibly in a stratified rock, resulting in the development of a mushroom-shaped (with a dome) intrusive in the host rock, it is termed as a laccolith (Fig. 1.3). This is the case with intrusions of viscous material that cannot spread too far sideways. In folded rocks, if the intrusion takes place at a later stage, it occupies the openings at the crest (in the case of anticlines) and trough (in the case of synclines) of folds. The resulting form of the intrusive of a curved-type is denoted as a phaccolith (Fig. 1.4). Large igneous intrusion of several kilometers in extent having a form which is convex upwards and concave downwards, is known as a lopolith (Fig. 1.5). Lopoliths are generally composed of gabbros.

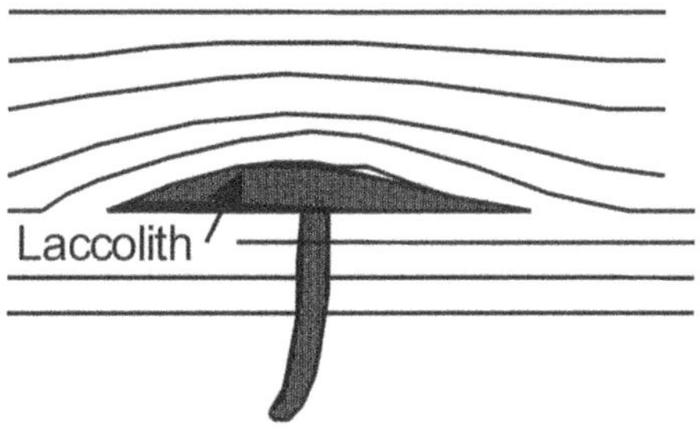

Fig. 1.3 Typical form of a laccolith.

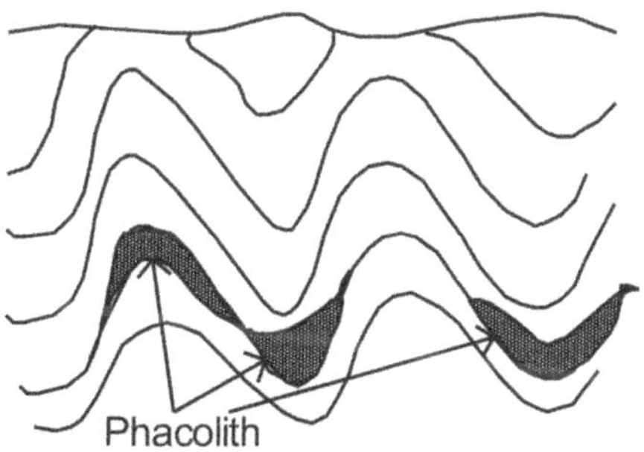

Fig. 1.4 Phaccoliths in folded rocks.

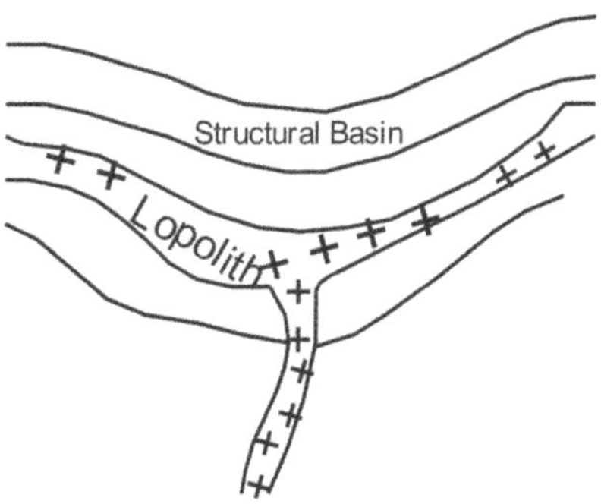

Fig. 1.5 Lopolith.

Very large intrusive bodies with steep outward slopes extending to great depths are known as batholiths (Fig. 1.6). Batholiths are composed of granites. Quite often, the base is not visible. The batholiths are present in orogenic regions and are associated with mountain building. The offshoot of a batholith, irregular in shape, is known as a stock while an offshoot having a circular pattern is called a boss.

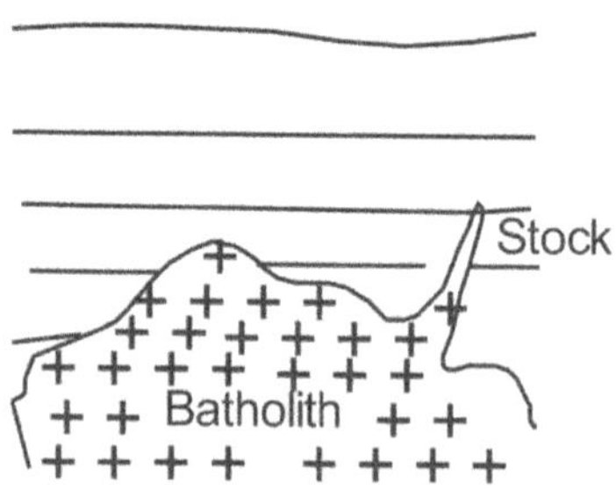

Fig. 1.6 Batholith.

Sedimentary Rocks

The source material for these rocks is contributed by weathering of the pre-existing rocks. The nature of the material derived from weathering depends on the type of weathering. Mechanical weathering contributes to the clastic or detrital load, which is in the form of disintegrated material of small sizes. Chemical weathering on the other hand involves dissolution of material. The product of weathering in such a case is in the form of dissolved load. As already indicated in chapter 3, removal of material from the site of weathering is through transport of both the suspended (clastic) load and the dissolved load during erosion. Erosion involves weathering and transportation of material.

Formation of sedimentary rocks involves deposition of the clastic and chemical sediments, lithification (compaction) and cementation of the particles with matrix resulting in a solid sedimentary rock. As indicated in chapter 3, deposition of clastic material is controlled by the change in physical parameters of the transporting agency while the deposition of the chemical load is a result of variations in the chemical framework of the system. Deposition can be under continental (fluvial, laccustrine, glacial or eolian), marine (in the ocean at different depths) or in the transitional (deltas and near–shore) environments. The depositional environment of a sedimentary rock is reflected in the physical, mineralogical and structural characteristics of the rock.

Classification of Sedimentary Rocks

Sedimentary rocks are classified under two broad divisions depending on their formation from detrital or Chemical load.

The clastic or detrital rocks are composed of the clastic particles. Nomenclature of the rocks in this category is based on the grain size of the particles.

Clastic material is classified on the basis of grain size ranges as follows :

	Particle size (mm)
Rudaceous	
Boulders	>256
Cobbles	64-256
Pebbles	4-64
Granule	2-4
Arenaceous	
Sand	1/16 -2
Silt	1/256-1/16
Argillaceous	
Clay	<1/256

Rock Types

Clastic sedimentary rocks

1. ***Boulder bed :*** This is formed under glacial environment. The load carried by a glacier is assorted in its sizes. The material is of different size ranges comprising of the boulder, cobble and pebble sizes. Their composition can also be varied depending on the source rocks. All these rock fragments are cemented by a matrix. Presence of a boulder bed in a geological sequence indicates prevalence of glacial conditions during the deposition of the sediments.

2. ***Conglomerate :*** This rock type is characterized by the presence of rounded pebbles of rock material cemented together by a fine-grained matrix. These pebbles are derived by weathering from the underlying formations in geological sequence. If a conglomerate has only pebbles of single mineral like that of quartz or chert, it is known as oligomictic conglomerate. On the other hand, if the pebbles are of diverse mineral composition, it is known as polymictic conglomerate. Presence of conglomerate in a geological sequence signifies period of non-deposition during which time weathering was active. It is an evidence for the existence of an unconformity.

3. ***Breccia :*** The rock fragments are similar in size as in a conglomerate but are angular in shape instead of being rounded. Breccia has broken angular fragments of rock material cemented by a matrix. Breccia occurs in fault zones where the rock deformation has resulted in the fragmentation of the rock.

4. ***Sandstone :*** The rock has grains in the sand-size range (2mm to 1/16 mm). Within this range, depending on the nature of grain size in the rock (whether towards the coarser side or towards the finer side), the sandstone is accordingly designated as coarse-grained or fine-grained sandstone. The matrix is of silica (arenaceous), iron oxide (ferruginous) or carbonate (calcareous) composition.

On the basis of the composition of the mineral constituents within sandstone, the following subdivision is made :

Ortho-quartzite (also called as quartz-arenite) : Dominantly of quartz with some chert. It contains often only quartz. Its purity in terms of only quartz indicates sufficient duration of weathering for the elimination of other susceptible minerals. The sandstones of the Cuddapah and Kurnool sequences belong to this category.

Arkose : Composed of quartz with feldspar (around 25% or more), it is formed by deposition of material derived from granites or quartzo-feldspathic rocks on weathering. Presence of feldspar, which is otherwise susceptible for weathering, indicates quick erosion after insufficient weathering of parent rock.

Graywacke : It is composed of quartz, feldspar and ferromagnesian mineral particles. Because of the presence of the dark mineral grains in considerable amount, the rock has a greyish colour in appearance. The source rocks for a graywacke are basic in composition. The sandstone formations in the Gondwana sequence are of graywacke type.

5. ***Siltstone :*** This rock is composed of silt-sized particles (0.06mm-.004mm).
6. ***Shale :*** Composed dominantly of clay minerals, shales exhibit layered (bedded) structure. If Ca- or Mg- carbonate minerals are present, the shale is calcareous or dolimitic shale.
7. ***Mudstone :*** Compact in appearance, this formation consists of clay as the dominant fraction together with fine particles of feldspar, quartz and sometimes carbonate.
8. ***Claystone :*** This rock is composed entirely of clay. Claystone is present in the Siwalik sequence at the foundation site of the Bhakra dam.

Sedimentary rocks of Chemical Origin

9. ***Limestone :*** This is composed of calcite (Ca-carbonate). If dolomite (Ca-Mg-carbonate) is present, it is designated as a dolomitic limestone. The limestones in the Cuddapah sequence (Vempalli limestones) are dolomitic in nature.

   Carbonate minerals in a limestone can be inorganic (through precipitation from solution) or organic (from chemical secretion by organisms) in nature. Fossiliferous limestone, containing relicts of fossil organisms, are biogenic in origin

10. ***Evaporites :*** These are formed from lake or sea water by evaporation. Gypsum and halite deposits are of this type.
11. ***Chert :*** Chert is a chemical precipitate of silica. Although considered under the sedimentary rocks, chert is often a secondary mineral formed from silica solutions and occurs as nodules, concretions and bands in sedimentary rocks. Narji limestones in the Kurnool supergroup within the Cuddapah Basin in Andhra Pradesh contain lenses of chert.

Sedimentary Structures

Several primary structures are evidenced in sedimentary rocks. These structures offer significant evidences of the depositional conditions (environments). These are primary in the sense that they are related to the mode of formation of the rocks and not developed subsequently. Tectonic structures, due to deformation, belong to the secondary category.

Stratification or Layering : Strata (individual layers) vary in thickness, indicating the time period involved in their deposition. Variation in composition in different strata is seen as colour change in individual layers. If undisturbed during or after deposition, the strata are horizontal.

Cross-bedding : Strata are inclined to each other. Cross-bedding is evidenced in shallow-water deposits or eolian (wind-borne) deposits. A typical cross-bedding feature exhibits horizontal bottom strata followed upwards by a set of inclined strata and again a set of near-horizontal strata on the top. The middle set (inclined strata) are asymptotic to the bottom set and are truncated at the top (at the contact with the top set) (Fig.1.7). The inclination of the middle set is towards the current direction. By noting the current directions as exhibited in the field by the sandstones at different places, it is possible to infer the paleocurrent directions during the deposition of sediments in a region. Once the current directions are inferred, the source material that contributed to the formation of these rocks can also be inferred. The location of the source material is towards the direction (upstream) opposite to that of the current direction.

Since the truncated part of the middle set is towards the top in a sequence, current bedding is useful for inferring the top direction of a geological sequence in disturbed areas. For example, using the top and bottom criteria from the cross-bedding in rocks with isoclinal folding, the anticlinal or a synclinal nature of the fold can be established.

Fig. 1.7 Cross-bedding in Navajo sandstone, Zion National Park,Utah (U.S.A.) (Courtesy: U.S. Geological Survey Photo: E.T.Nichols).

The type of cross-bedding is indicative of the natural agency involved in deposition. In eolian deposits, asymptotic trend with the bottom set is not present. This type of cross-bedding is known as tabular cross-bedding. It has truncated features with both the top and bottom sets. In zones of agitation, like in a beach environment, the cross-bedding shows frequent change in the direction of the current. This is termed as an agitated current-bedding or festoon type (Fig. 1.7).

Graded Bedding : This structure is typical in relatively deeper and quiet water. The particles get sorted as they settle down forming a gradation of size from bottom to the top of the beds. Coarser particles settle down first followed by particles of relatively smaller dimensions progressively to the top (Fig. 1.8). Graded bedding is associated generally with graywackes and is related to deposition of the material carried by turbidity currents. Using graded bedding, the top direction in a geological sequence can be established. In each unit of graded bedding, the finer particles are towards the top.

Fig. 1.8 Graded bedding (King George IV Lake Area, New Foundland) (Courtesy: Geol. Survey, Govt. of New Foundland & Labrador).

Ripple Marks : Ripple marks are characteristic of shallow water deposition and also of eolian deposits. Ripple marks are of different types – symmetric or asymmetric ripples. Symmetric ripple marks are characteristic of still water or a surf zone. Asymmetric ripple marks (also known as current ripples) have typical geometry with the stoss side followed by a lee side. The stoss side has a gentle slope compared to the lee side. Crest is the highest point separating both sides. The direction from stoss to the lee side indicates the paleocurrent direction. Ripple marks in fluvial and eolian deposits are of the asymmetric type. Just as the agitated current bedding, superimposed ripple marks are generated in an agitated near-shore situation.

Ripple marks vary in their nature depending on the hydrodynamic conditions of the stream and the associated environmental framework in a depositional basin. Ripple marks in Kurnool supergroup within Cuddapah Basin provide a good example to this effect. In this basin, the sequence of Kurnool supergroup is essentially an orthoquartzite-limestone-shale type deposited in two cycles. Above the basal conglomerate in the sequence is the Banaganapalle orthoquartzite. This formation exhibits different types of ripple marks at different locations in the Basin. The schematic representation of different types along with the location index is presented in Fig. 1.9 .In general these ripples are symmetrical in nature. Their branching and sinuous nature is indicative of their current origin. Based on these ripple types, the depth and flow conditions for the deposition of these orthoquartzites have been inferred. Their general north-south trend with the stoss side to the west denotes the current flow from west to east. These ripples are indicative of a low-energy near-shore environment (Rao and Gokhale, 1973). The sinuous and cross-ripple types are reported to be formed at very shallow depths under high velocity. The straight ripples are indicative of an increase of depth and a reduction in the current velocity in the depositional basin.

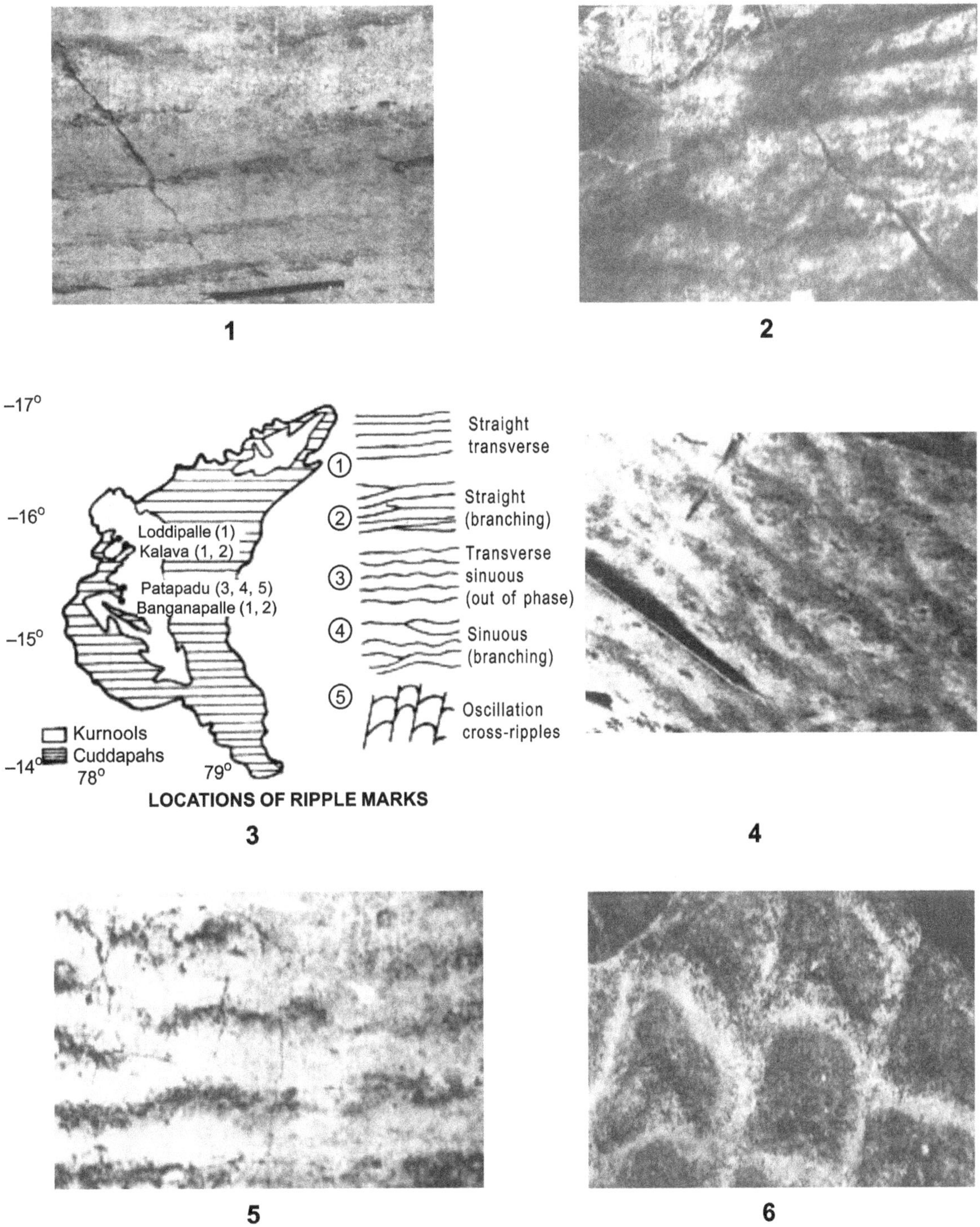

Fig. 1.9 Different types of ripple marks in the Kurnool formations within the Cuddapah Basin (Rao and Gokhale, 1973).

Mud Cracks : Typical of argillaceous (clay) sediments, these are polygonal shrinkage cracks (Fig. 1.10) generated during the drying of a sediment. They indicate sub-aerial exposure of sediments before lithification.

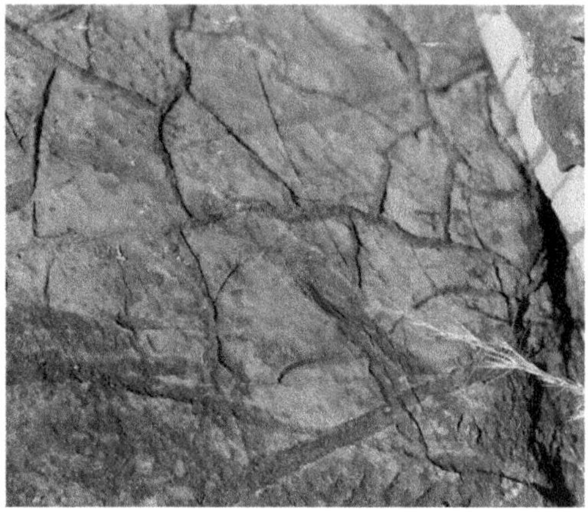

Fig. 1.10 Mud cracks in Vempalle formations, Cuddapah Basin (Photo: G.S.Sukumar).

Sedimentary Environments

The environmental conditions for the deposition of a group of sedimentary rocks are reflected in the type of formations, sedimentary structures and the geological sequence. For example, presence of cross- bedding and ripple marks is indicative of shallow depth conditions. An orthoquartzite (quartz-arenite) – shale- sequence indicates a transgressive marine environment. In such a geological sequence, the thickness of the shale horizon is less compared to the quartz-arenite or the limestone. Occurrence of more than one cycle of sandstone-shale-limestone series in a geological sequence is indicative of periodic changes in the depth of the depositional basin. The geological sequence of Cuddapah and Kurnool supergroups in the Cuddapah Basin or the sequence of formations in the Vindhyan Supergroup is example of the same.

As already explained, if the sequence starts with a boulder bed, it is indicative of glacial environment. The Siwalik formations or the Gondwana formations have the boulder bed in their sequence. In the Gondwana sequence, presence of varved clay is indicative of deposition in a fresh water glacio-laccustrine environment. The laminations of the clay reflect the seasonal changes. Presence of coal seams denotes a reducing environment favouring decomposition of vegetation.

If a geological sequence is only of sandstone and shale with no limestone, paucity of limestone is indicative of the chemical conditions of the depositional environment with pH range not favourable for the precipitation of carbonate. Limestone deposition is only in high pH conditions (around pH of 8.0). The other alternative explanation for the absence of limestone is that no source material with carbonate load was available for deposition. Typical example of sandstone-shale sequence without limestone is in the coal basins within the Gondwana sequence.

Presence of graywacke type of sandstone and typical graded bedding structure are indicative of relatively deeper and stable water.

Reworked Sediments : Sedimentary rocks can also be formed from source material derived from other older sedimentary formations from the same sequence or from other locations. Thus the source material undergoes through another round of weathering and erosion. This results in the removal of all susceptible minerals and also in the rounded shape of the particles due to further erosion. The sequence in terms of the degree of susceptibility of minerals to weathering conforms to the Bowen's Reaction Series. Quartz is the least susceptible (most resistant) while pyroxene and amphibole are most susceptible. Mineral structure and composition control its stability during weathering and erosion. The greater the number of anions surrounding a cation in the mineral structure, the greater is its instability. Mineral hardness also plays a role in the particle breakdown.

Diagenesis : This process involves post-depositional changes under compaction or burial of sediments. As already indicated, the diagenetic process involves very low T and P conditions. During diagenesis, dissolution and recrystallization features are evidenced in the form of stylolites (Fig. 1.11) in limestones and authigenic overgrowths on individual minerals in the sedimentary rocks. These are due to the pressure phenomena.

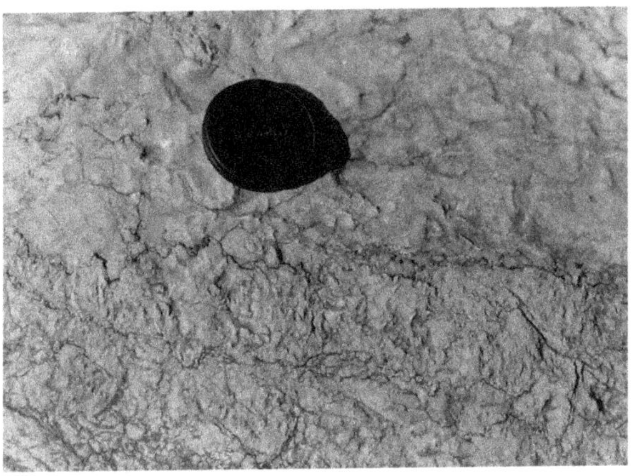

Fig. 1.11 Stylolites in vempalle limestones (Cuddapah Basin) (Photo: G.S.Sukumar)

Localized deformational features in terms of contortions (folding) and minor faulting in the formations are also a result of diagenesis. These are termed as penecontemporaneous deformational features (being contemporaneous with deposition).

Metamorphic Rocks

These are formed through the transformation of the pre-existing rocks under increased temperature and pressure conditions. This process of transformation is known as metamorphism. The changes that take place above the temperature (T) and pressure (P) limits of about 150 °C and 300MPa respectively are included under metamorphism. However, the post-depositional modifications in sedimentary formations due to compaction and burial are below these T and P ranges and hence are not included under the metamorphic processes. They are ascribed to diagenesis.

Under new T and P conditions, the earlier minerals in the rock are unstable. Consequently, new minerals form through recrystallization in the rock. This recrystallization process is said to take place under solid-state conditions in which the ions migrate along grain boundaries and mineral lattices.

Formation of a metamorphic rock from a pre-exiting igneous or sedimentary rock is controlled by the following parameters:

1. ***Composition of the rock*** : The overall chemical constituents in the rock will remain same. However, the mineralogy will alter suiting to the new environment, as dictated by the new T and P conditions.

2. ***Temperature*** : Temperature increases with depth due to the geothermal gradient. In general, this increase is at 30 °C/km. This is the common source of heat in metamorphism taking place at different depths. The heat also can be supplied by magma. For example, around an igneous intrusive (dyke or a sill), the host rock is altered in the vicinity of its contact due to high temperature. In this case, the metamorphism is confined to the contacts of the rock with the igneous intrusive.

   Increase of temperature results in the increased activation energy for the reactions taking place during metamorphism. As already indicated, new mineral assemblage is formed on recrystallization. The volatile constituents often escape during the process.

3. ***Pressure*** : Pressure increases with depth. This pressure, known as the confining or the hydrostatic pressure, is equal in all the directions. In the stress terminology, the longitudinal stress is equal to lateral stresses ($\sigma_1 = \sigma_2 = \sigma_3$). Under these conditions, since a mineral is compressed from all directions during its formation, the density of the new mineral improves. Thus, in general, high-density minerals form under high pressure. The mineral growth takes place uniformly in all the directions.

   The pressure can also be unequal in different directions. This is true in the case of deformation or tectonic activity where the stress in one direction exceeds over the same in the other two directions. This is known as differential stress condition. Under such conditions, the minerals in the rock during recrystalliztion grow with preferred orientation. This growth is in the direction of the least principal stress.

   The structure in a metamorphic rock is thus controlled by the stress conditions during its recrystallization. The metamorphic rocks can be categorized on the basis of structure as follows :

   Foliated : These are formed under differential stress conditions. They exhibit planar structure in the form of gneissose structure (the corresponding rock is known as a gneiss), schistosity (the corresponding rock is termed as a schist) or cleavage (as in slate). Planar structure develops when the stress in the longitudinal direction (σ_1) is greater than the other two stresses ($\sigma_2 = \sigma_3$) Foliation thus is due to preferred orientation of silicate minerals such as clays.

If the longitudinal stress is equal to one of the lateral stresses ($\sigma_1 = \sigma_2$) but exceeds the other one (σ_3), a linear structure (lineation) develops with growth of minerals in a linear fashion.

If the stresses in all the three principal stress directions are unequal ($\sigma_1 > \sigma_2 > \sigma_3$), both the planar and linear structures form.

Non-foliated : These are the metamorphic rocks formed under hydrostatic stress conditions with minerals growing without any preferred orientation. Examples of this type are quartzite and marble.

4. ***Chemically active fluids :*** The most common fluid is water with dissolved constituents. These fluids facilitate the ion migration or even act as catalysts in controlling the reaction rates.

Types of Metamorphism

1. ***Contact metamorphism (Thermal metamorphism) :*** Temperature is dominant with generally low-pressure. As already explained, this type is prevalent at the contacts of rocks with igneous intrusives The degree of alteration in the host rocks decreases with increasing distance from its contact with intrusive. The altered zone in the rock is termed as a contact aureole. The temperature of the intrusive, composition of the intrusive material and composition of the host rock control the composition of the altered material. If the intrusive reacts with the host rock, addition of the constituents from the intrusive results in the formation of new minerals. In such cases the term contact metasomatism is used. For example, in the Cuddapah Basin in Andhra Pradesh, intrusion of sills in the limestone has resulted in the formation of barytes, steatite and asbestos deposits in the vicinity of the contact of the sill with the host rock.

   If the heat source is a volcano, the term pyro- metamorphism is used

2. ***Dynamic metamorphism :*** The process is characterized by the dominance of pressure. The rock exhibits evidences of deformation effects. Differential stress is the dominant factor. In fault zones, due to the stresses induced through displacement of formations, intense crushing and deformation features develop. The metamorphism in faulted zones is termed as cataclastic metamorphism.

3. ***Dynamo-thermal metamorphism :*** Both T and P are dominant. This process extends over large regions and hence called regional metamorphism. Rocks formed are varied (like gneisses, schists, marbles and quartzites) depending on the source rocks.

4. ***Retrograde metamorphism :*** This is operative in areas of tectonic uplift. With the decrease of depth of burial, development of relatively low T-P mineral assemblage takes place in the already metamorphosed rocks. Consequently, changes take place in the reverse direction. However, retrograde metamorphism is not very common. This is due to several reasons. Hydrous minerals cannot form since volatiles are already expelled earlier in these metamorphic rocks. The reaction rates are very slow due to the absence of volatiles which act as catalysts. Reactions are very slow also because of low temperatures at which the diffusion is also at a reduced pace.

Classification of Metamorphic Rocks

Non-foliated

1. Quartzite :
   - characterized by the presence of interlocking grains of quartz
   - formed from a sandstone with high quartz content
   - due to recrystallization, the matrix (cementing material) in the source rock (sandstone) is nonexistent . Thus the resulting rock is compact , massive and hard.
   - minerals have sutured outlines
2. Marble :
   - formed from limestone
   - composed of calcite (and dolomite if Mg present) exhibiting interlocking grain fabric
3. Hornfels :
   - formed from shale
   - common in contact metamorphism (with dominant temperature)
   - exhibits equigranular growth
4. Granulite :
   - formed under high T and P conditions
   - minerals like mica and other sheet silicates are unstable and so foliation is absent in high grade metamorphic rocks
   - minerals like garnet and pyroxene impart a granular texture
5. Eclogite :
   - formed from basic igneous rocks under high T-P conditions.
   - characterized by the presence of garnet and pyroxene (omphacite)

Foliated

1. Slate :
   - characterized by closely spaced parallel partings (cleavage)
   - formed from shale under low metamorphic conditions
2. Phyllite :
   - characterized by foliation
   - presence of mica imparts a silky and smooth surface
   - forms from slate under increasing intensity of mrtamorphism
3. Schist :
   - formed from either a phyllite or an igneous rock
   - characterized by irregular but distinct foliation (schistosity) since the grain size gets larger. While minerals like quartz and feldspar do not impart foliation, sheet silicate minerals such as micas contribute to the foliation. Depending on the dominant planar mineral, the schist is differentiated as a mica-schist, chlorite-schist, graphite-schist etc.

4. Gneiss :
   - source rocks are mostly granites or quartzo-feldspathic rocks.
   - characterized by gneissose structure (alternate layers of of dark and light minerals). The sheet silicates are unstable. Quartz, feldspar and dark minerals such as amphibole are the main constituents. However, their preferred orientation exists.
   - the alternate dark and light- coloured layers are due to differentiation (separation of acidic and basic minerals)
   - just as in the case of a schist, a gneiss is differentiated as a feldspathic gneiss, amphibole gneiss etc.
5. Amphibolite :
   - dark coloured rock with plagioclase and hornblende
   - formed from basic or intermediate igneous rocks
   - foliation varies. It is pronounced if planar minerals are present.

Khondalite belongs to the granulite category and is composed of garnet and sillimanite. The rock is named after the hill tribes (khonds). The Easten Ghat ranges are composed of this rock type. It contains graphite. The graphite deposits of India are associated with this rock type.

Charnockite is a typical rock type occurring in S.India. The rock has been named in honour of Job Charnock, the founder of Calcutta City. Charnockite is characterized by the presence of hypersthene (pyroxene). The rock is coarse-grained. The rock, although considered by some as an igneous type, is essentially metamorphic with the mineral assemblage of a granulite (high grade metamorphic rock).

Intensity of Metamorphism

Metamorphic status of a rock, as a function of temperature and pressure, is indicated in terms of metamorphic grade and metamorphic facies.

Metamorphic grade : Index minerals are used for this purpose as indicators of a specific T and P range. These are characteristic minerals, which are stable in specific ranges of pressure-temperature. First appearance of an index mineral is taken as the starting of a zone of particular T-P conditions pertaining to that mineral. In an area, all such locations are connected through a line called as an isograd. Hence an isograd defines locations with identical temperature and pressure of metamorphism. Areas between different isograds are marked as different zones. Through fieldwork and study of minerals in different metamorphic rocks, a geological map with different metamorphic zones is prepared.

Typical index minerals, in the increasing order of the metamorphic grade, are chlorite, biotite, garnet, kyanite and sillimanite. Accordingly the zones are termed as biotite zone, chlorite zone etc. (Fig. 1.12).

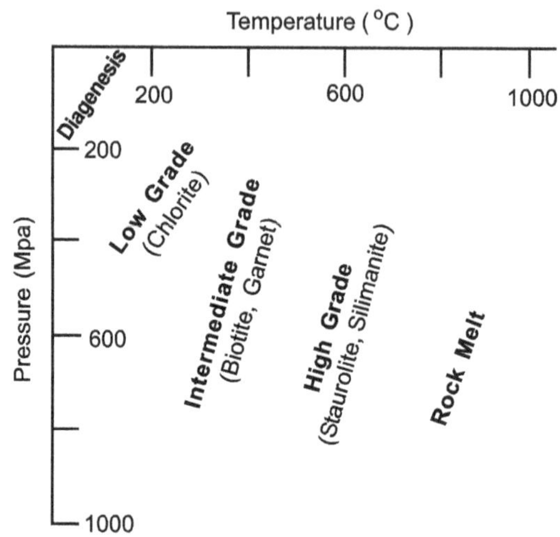

Fig. 1.12 Generalized Pressure-temperature diagram for metamorphic grades.

Metamorphic facies : The metamorphic facies concept uses a mineral assemblage (a group of minerals) which is stable under a definite T-P range as indicative of a particular facies. This is regardless of the rock types in the region. As long as the mineral assemblage of a particular species is identified in an area, the area is classified into a specific metamorphic facies regardless of the variation of rock types in the same.

The boundaries for different facies as a function of T and P are indicated in Fig.1.13.

Each of the facies indicated is characterized by a group of stable minerals under the specific T and P conditions.

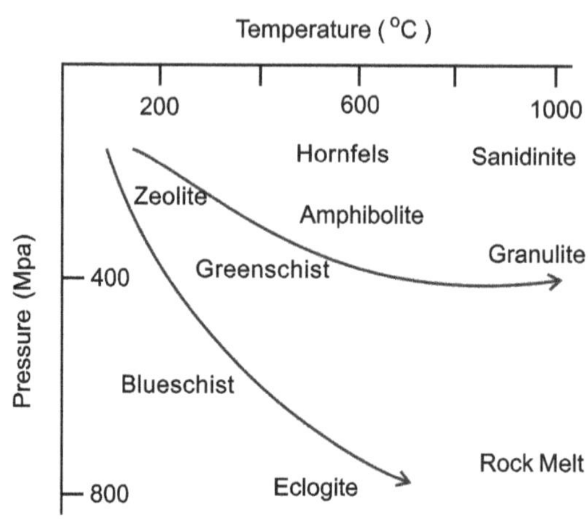

Fig. 1.13 Generalized picture of metamorphic facies as a function of temperature and pressure.

For example, a green schist facies is characterized by chlorite and amphibole. The granulite facies is marked by the presence of pyroxene and garnet. Charnockite, typical rock type in the Archaens of S.India, is composed dominantly of garnet and hypersthene (a pyroxene) and is relegated to the granulite facies. The hornfels facies conforms to high temperature and low pressure conditions. As already indicated, rocks in the contact metamorphic regions pertain to this facies.

It may be noted that a normal geothermal gradient (at 30 °C/km increase with depth) passes through zeolite – green schist – amphibolite - granulite facies from low T-P to high T-P conditions. On the other hand, if the geothermal gradient is steeper in a region, it is in the direction of zeolite – blue schist – eclogite facies with increasing metamorphic status.

As an example, metamorphism of a basic igneous rock (gabbro) can be considered. Gabbro has plagioclase feldspar and pyroxene as essential minerals. If the rock is metamorphosed to green schist facies (under relatively low T and P conditions), amphibole and chlorite develop as characteristic minerals. With increasing metamorphic status to a granulite facies, this rock (gabbro) has pyroxene and garnet as the stable mineral assemblage.

Rocks as Construction Materials

Rocks find extensive use in a variety of construction operations ranging from buildings to dams, highways and railroad ballast. Besides these direct uses, they are also used in the form of aggregates in cement concrete.

Granite, sandstone, slate and marble are commonly used as building stones, in the form of blocks, cut slabs or sheets. For this purpose, they should be resistant to weathering (durable), strong (with high compressive as well as transverse strength values), and should be of uniform mineralogical composition. As decorative stones, they should be free from surface flaws and take a good surface polish. Granite is good if it is compact and without any closely-spaced fractures. Granites take good polish and are available in different textures and colours.

Slate has a high tensile strength and finds use as roofing or topping material, especially in Himachal Pradesh In the case of sandstone, composition of the cementing material (matrix) is important. It should be siliceous (composed of silica) instead of a ferruginous (iron oxide) or a carbonate (calcite, dolomite) in composition. Silica is highly resistant to weathering.

Marble is used in a variety of uses. Being metamorphic in nature, it has already withstood pressure-temperature conditions in nature. It has been in use in India for the construction of palaces, temples and other historical monuments. It has been increasingly used in the construction of buildings of religious interest. Besides, marble flooring is very popular. Marbles come in a variety of colours, textures and compositions.

In India, granite, sandstone and marble have been extensively used in the historical past for construction of palaces, monuments, temples and other religious shrines The temples of Mahabalipuram, Konark, Kajuraho and the ancient temples of South India are but few examples of stone architecture From the Moghul period (with monuments like Taj Mahal) till the present day (Lotus temple in New Delhi and the Dayalbagh temple near Agra), marble has been increasingly in use as a building material. The building and decorative stone industry in India has developed rapidly over the years and is currently supplying the material to several countries.

There is a tremendous improvement in the exports too. While granite blocks and natural cleft stones of sandstones, limestones, quartzite and slate were the exports around two decades ago, with improvements in sawing, polishing and cutting processes the exports in recent times are in the form of slabs and polished tiles. India ranks third in the World market in regard to granite exports (around 9.5%) with China and Italy occupying the first and second positions in the World market with 19.5% and 14.7% respectively.

Distribution of Building and Decorative Stones in India

Granite

India has around 1690 Mcum of reserves, amounting to around 20% of the World's reserves (Alok Kumar and Kulveer Singh, 2003). More than 200 shades are available. With granite occurrences in 17 districts, Andhra Pradesh is one the largest producers of granite in the country. Several varieties like Black Galaxy, Jet Black, Blue and White are very popular abroad. Many of the varieties are from Prakasam, Guntur, Warangal, Khammam and Srikakulam districts

With well-developed infrastructure for granite industry, Tamil Nadu is a leading producer of granite. Among the many occurrences, deposits in Dharmapuri, Madurai, Salem, Coimbatore and Kanyakumari districts are important sources of supply.

In Karnataka, the Peninsular granites and closepet granites are very popular as decorative stones. Deposits in Bijapur, Bangalore, Raichur and Mysore districts supply several varieties with pleasing colours.

In Orissa, occurrences are in Ganjam, Koraput, Bolangir and Kalahandi districts. The red granite (Bundelkhand granite) from Lalitpur and Jhansi districts of U.P. is very popular.

Marble

Of the 1200 M tons of estimated Indian reserves, around 90% are in Rajasthan. The occurrences are distributed mainly in Nagaur, Udaipur, Jaipur, Ajmer, Alwar, Rajsamand, Bhilwara,Sirohi and Dungarpur districts. The Makrana marble has been used in several monuments including Taj Mahal . Marble producing centres from Makrana, Udaipur, Banswara , Rajsamand, Udaipur and Jaipur contribute to major supplies from the state.

Marble occurrences in A.P. are mainly located in Khammam , Guntur and Cuddapah districts. In Gujarat, white and black marble varieties occur in the Ambaji region. Marble occurrences in M.P. are in Katni and Jabalpur districts.

Sandstone

Over 90% of the Indian deposits are in Rajasthan mainly occurring in Bharatpur, Dholpur, Kota, Jodhpur, Sawai-Madhopur, Bundi and Chittorgarh districts.Several shades including pink, brown, red and yellow are available. Several monuments in India are constructed with the red sandstone from Rajasthan, Besides, sandstone occurrences are reported from several states including Assam, Meghalaya, Bihar, Gujarat, A.P., Karnataka, Tamilnadu and West Bengal.

Flaggy Limestone

Important occurrences are in A.P., Karnataka and Rajasthan. In A.P., the deposits are situated in Ranga Reddy, Nalgonda, Mahaboobnagar, Cuddapah, Anantapur and Kurnool districts. In Rajasthan, it occurs in Chittorgarh and Jaisalmer districts.

Natural Concrete Aggregates

Natural aggregates include rocks, gravel and sand. Aggregates amount to about 70% by volume of the concrete. Geological considerations play an important role in the choice of aggregates. Mineral composition, texture and impurities are important in the material selection. A good aggregate material should be durable, clean without any impurities, uniform in quality as well as size, capable of providing an efficient bond with the cement, weather-resisting and should not react with the cement in the mix. Presence of impurities in aggregates is undesirable as they retard the hydration, prevent adhesion with the cement, lower the strength and decrease the durability. In India, granite, quartzite, basalt, sandstone, and limestone are commonly used as rock aggregates. River sand is used as fine aggregate.

Alkali-aggregate Reactions: On hydration in a cement mix, $Ca(OH)_2$ is released as a product of the hydration reactions. Consequently, the pH of the system increases resulting in an alkaline environment. Aggregates in the concrete should be stable under such conditions. This requires the absence of any deleterious or reacting component in the aggregate composition .The reactions are essentially expansive. Presence of considerable moisture and high temperature enhance the reaction.

Alkali-aggregate reactions can be of alkali-silica, alkali-silicate and alkali-carbonate types (Gillott, 1975). The alkali-silica reactions are due to the presence of highly reactive colloidal and microcrystalline silica minerals such as chert, opal and volcanic glass. These constituents form a gel on dissolution, resulting in an increase in the volume of the system. The alkali-silicate reactions take place if the aggregates contain minerals like talc, chlorite, kaolinite, muscovite and serpentine. These minerals enter into complex reactions in an alkaline environment l leading to a considerable increase in volume. The alkali-carbonate reactions are common to dolomitic limestone minerals. Dolomite in converted to calcite (dedolomitization) accompanied by an increase in volume. Limestone aggregates are not considered bad as the calcite grains develop reaction rims, which serve as effective bonding sites in the concrete. Thus these reaction rims are beneficial in alkali- aggregate reactions (Hadley, 1964; Mather, 1974).

Chlorides and sulphates are deleterious to concrete. The aggregates should not contain these constituents. Chloride is a source of corrosion in concrete. The sulphates convert the free lime in the concrete to calcium sulphate leading to a substantial increase in volume finally resulting in the cracking of concrete. Pyrite and marcasite (iron sulphide minerals) are undesirable in aggregates as they oxidize leading to the decrease in strength of the concrete. The iron present causes staining of the concrete.

2

Engineering Properties and Classification of Rock Masses

Rock masses are composed of different minerals of varying strength and susceptibility to weathering. During deformation, various structural features develop in the rocks. These include joints, lineaments, folds and faults. The weak planes developed in the rock masses facilitate weathering through the action of natural agencies such as water. Thus the engineering behaviour of a rock mass is controlled largely by the joints or discontinuities, in addition to its strength and deformational characteristics. Reliable data from field tests on in situ rocks and laboratory tests on samples are very essential for ground reinforcement during dam construction, landslide prevention, seepage control and appropriate tunnel linings and rock supports.

A. Strength Characteristics

1. *Unconfined compressive strength (UCS)*

The core sample is subjected to loading in the vertical direction (uniaxial loading). For this purpose, cylindrical core specimens chosen should have length of at least twice their diameter. However, it is advisable to keep the length-diameter ratio less than 3. The ultimate failure load divided by the area of cross section of the specimen corresponds to its unconfined compressive strength.

The possible modes of failure in the specimen are indicated in Fig. 2.1 (a). The most common type is shear failure. Due to friction at the sample-platen contacts, radial stresses do not occur. The state of stress in different regions of the sample is indicated in the

Fig. 2.1 (a) If the friction is absolutely eliminated (which cannot be achieved), then the ideal tensile fractures occur. If the friction is eliminated partially by appropriate means during testing, a combination of both tensile and shear patterns of failure results in the specimen.

The ultimate failure strength of a specimen depends on its composition, fabric, structural features (such as bedding, cleavage, schistosity and gneissic banding) as well as its orientation to the direction of loading.

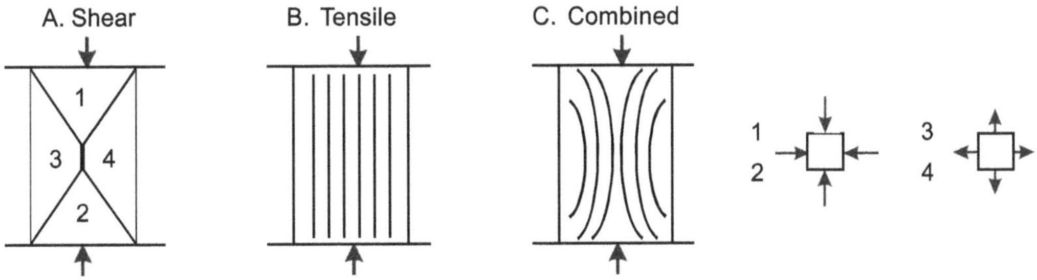

Fig. 2.1 (a) Typical failure patterns in a cylindrical specimen subjected to uniaxial compression.

Typical stress- strain patterns are indicated in Fig. 2.1 (b).

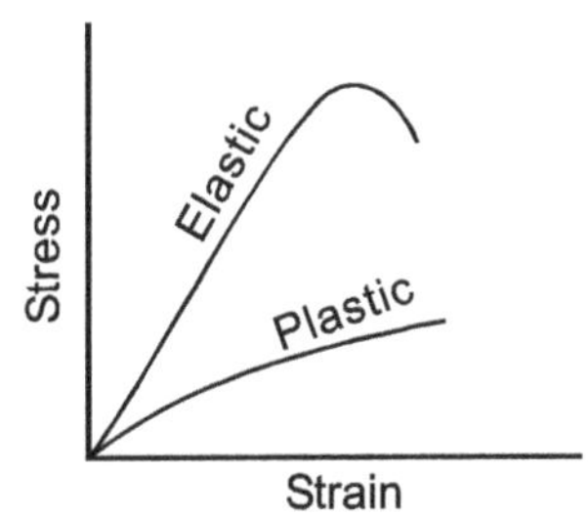

Fig. 2.1 (b) Typical stress-strain curves.

Unconfined compressive strength values for different rock types are listed in Table 2.1.

Table 2.1 Typical ranges of unconfined compressive strength

	kN/m^2	(MPa)
Granite	40,000-290,000	40-290
Basalt	180,000-275,000	180-275
Sandstone	10,000-230,000	10-230
Siltstone	25,000-250,000	25-250
Shale	6,500-200,000	6.5-200
Limestone	5,000-200,000	5-200
Quartzite	200,000-350,000	200-350
Gneiss	151,000-248,000	151-248
Schist	7,500-139,000	7.5-139
Slate	95,000-250,000	95-250
Marble	48,000-230,000	48-230

2. *Uniaxial tensile strength*

Conventional mode of tensile strength determination by gripping the specimen and pulling it apart is not feasible in the case of rock specimens since rocks are weak in tension. So an indirect method known as the Brazilian test is adopted.

The procedure involves testing on a disc, with its thickness of half its diameter, cut from a rock core. The specimen is loaded as indicated in the Fig. 2.2 (a). A compressive load is applied on the specimen in the manner shown in the figure. The tensile strength (σ_t) is given by

σ_t (in MPa) = 2F / (Π D L)

F = Load at failure

D = Diameter of the specimen (mm)

L = Length of the specimen (mm)

In this test, the pattern of failure is in the form of axial splitting. In addition, due to high stress concentrations at the contacts of the specimen with the loading plates, local crushing of the specimen is common. The stress distribution pattern is indicated in Figs. 2.2 (a) and 2.2 (b).

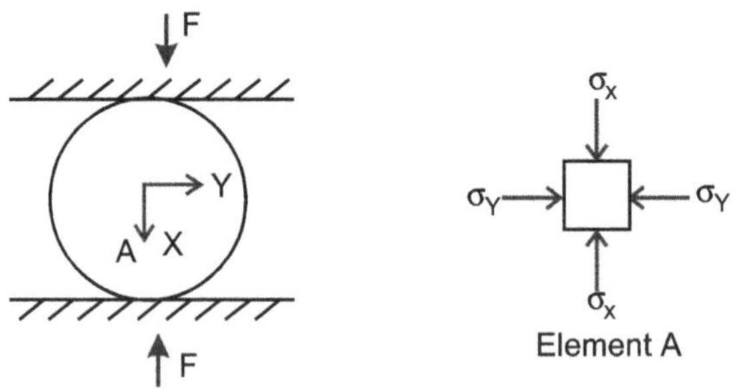

Fig. 2.2 (a) Loading of the specimen in indirect tensile strength test.

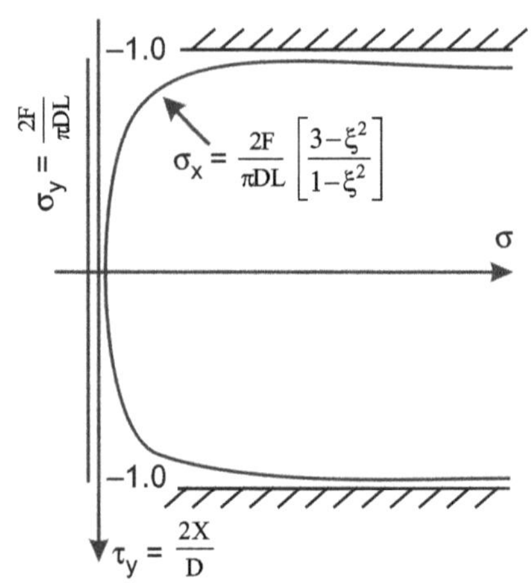

Fig. 2.2 (b) Theoretical stress distribution pattern In a specimen subjected to indirect tensile strength test (Gokhale and Rao,1981).

3. *Shear strength*

The specimen, confined under lateral pressure, is subjected to continuously increasing vertical compression. The specimen is housed in a triaxial cell and the lateral pressure (σ_3) is generated all around the specimen through the pressure cell assembly (Fig. 2.3).

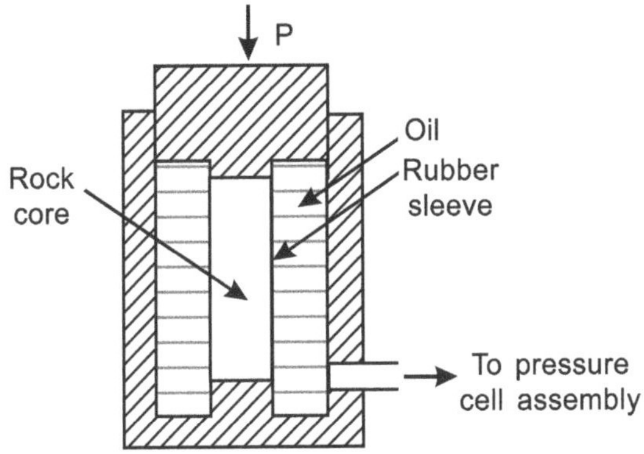

Fig. 2.3 Schematic section of a triaxial cell.

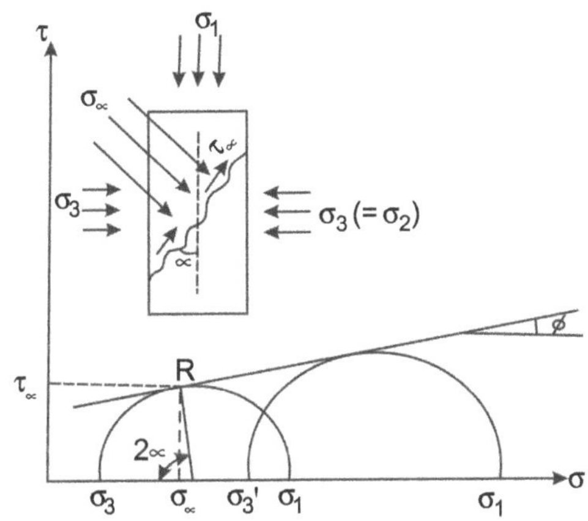

Fig. 2.4 Mohr diagram for estimation of shear parameters.

If the ultimate failure load is 'P ', the major principal stress (σ_1) is P/A, where A is the area of cross-section of the specimen. Using the stresses σ_1 and σ_3, the Mohr circle is drawn (Fig 2.4). A tangent drawn as in the figure located on the circumference of the circle by drawing the radius at an angle (twice the angle the failure plane makes with the vertical axis) represents the envelope.

The coordinates of the point R on the x- and y- axes provide the values of normal stress (acting at right angles to the shear plane) and the shear stress respectively. The angle the envelope makes with the horizontal is the angle of internal friction (φ).

It is easier to locate the envelope with test data on two or three samples. The envelope is then drawn tangential (Fig. 2.4) to all the circles.

B. Deformational Characteristics

4. The modulus of elasticity or the Young's modulus (E) and the shear or rigidity modulus (μ) are obtained from the stress-strain diagrams from the test data. The Young's modulus is defined as the ratio of the longitudinal stress to the longitudinal strain (σ_x / ε_x). This is used in the classification of rock masses. For this purpose, the tangent modulus at 50% of the failure load (Fig. 2.5) is commonly considered as E. These moduli values for a rock mass would enable an understanding of the ability of the rock mass to resist the respective stress conditions.

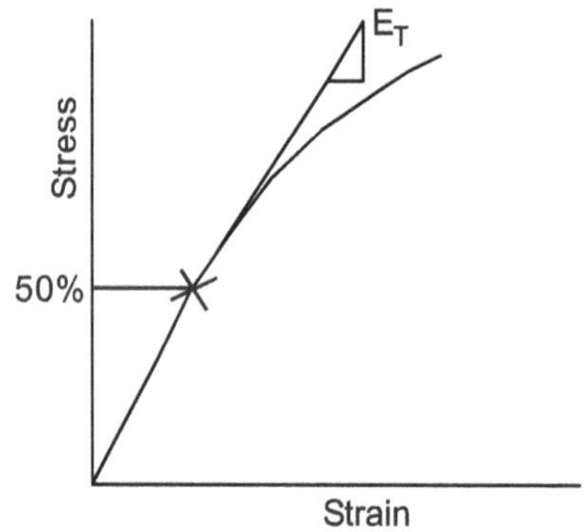

Fig. 2.5 Tangent modulus at 50% failure load.

Table 2.2 Ranges of modulus of elasticity for selected rock types

	kN/m^2 (x10^6)
Granite	10.35 - 82.11
Diorite	24.84 - 42.09
Basalt	40.71 - 85.56
Sandstone	4.14 - 55.20
Siltstone	6.90 - 64.17
Shale	2.07 - 68.31
Limestone	2.76 - 97.29
Quartzite	8.28 - 44.16
Gneiss	24.15 - 104.19

5. Poisson's ratio (ν) is the ratio of the lateral strain (ε_y) to the longitudinal strain (ε_x). The longitudinal strain refers to the shortening of the specimen due to the compression of the test specimen subjected to vertical loading. The accompanying change in the lateral direction provides the lateral strain. These strains are monitored through the strain transducers installed on the test specimen.

   Poisson's ratio (ν) is expressed in terms of modulus of elasticity (E) as

   $$\nu = \varepsilon y. E/\sigma_x$$

   Poisson's ratio for earth materials is around 0.3.

The elastic parameters obtained in the laboratory from the tests conducted on core samples are not representative of the in situ rock conditions. These are static measurements. In the field the in situ conditions include compositional changes, alterations due to weathering, joints or weak planes and moisture saturation.

Estimation of dynamic elastic parameters can be carried out on the rock masses in the field using seismic refraction method. This involves recording the travel-time values for longitudinal (P) and transverse (S) waves at different points along a traverse. The shock is generated through a shock hammer. Details of this procedure are indicated in chapter 6.

The elastic parameters can be expressed in terms of the P- and S- wave velocities as

$$E = (V_s^2 . \gamma / g) . [3 (V_p / V_s)^2 - 4] / [(V_p / V_s)^2 - 1]$$

$$\mu = (V_s)^2 . \gamma / g$$

$$\nu = 0.5 [(V_p / V_s)^2 - 2] / [(V_p / V_s)^2 - 1]$$

E = modulus of elasticity (kN/m^2)

μ = shear modulus (kN/m^2)

ν = Poisson's ratio

g = acceleration due to gravity ($9.8 m/s^2$)

γ = unit weight of the material (kN/m^3)

V_p = velocity of the P-wave (m/s)

V_s = velocity of the S-wave (m/s)

Engineering Classification of Rock Masses

The geological classification is based on the genesis or mode of origin of the rock and its composition together with its textural attributes. The nomenclature used enables a broad understanding of the features of a rock. But in the engineering practice, the strength and the deformation characteristics are important. The same rock type may exhibit different engineering properties at different locations. Besides composition, several other factors like the weathering status, joints, bedding planes, gneissose banding, schistocity, folds, faults and water saturation control the engineering behaviour of the rocks. Hence any classification intended for use by the civil engineers should necessarily have these attributes incorporated in the same.

In rock mechanics, the term rock mass is used. A rock mass can be defined as the rock formation in situ (at its site of occurrence) with all the associated features intact. Hence classification procedures evolved over the years in the past have been based on the rock mass characteristics. Rock mass classification contributes not only to an understanding of the strength and deformation properties of the in situ formations but also in the choice of tunnel supports.

1. *Classification based on strength and modulus :*

This classification proposed by Deere and Miller (1966) has been in use for over three decades and in fact has provided the basis for modification by subsequent workers.

In terms of the strength (unconfined compressive strength) and modulus (tangent modulus at 50% of the ultimate failure strength) as obtained from the core samples, rocks are classified as follows :

On the basis of strength

Class	Strength	Range , lb/in^2 *	(MPa)
A	Very high strength	> 32,000	(> 220.6)
B	High strength	16,000-32000	(110.3-220.6)
C	Medium strength	8,000 –16,000	(55.2 – 110.3)
D	Low strength	4,000 - 8,000	(27.6 - 55.2)
E	Very low strength	< 4,000	(< 27.6)

* units originally used in the classification

On the basis of modulus ratio (tangent modulus / ultimate compressive strength) :

Class	Modulus ratio	Range
H	High	> 500
M	Medium	200-500
L	Low	< 200

Based on the strength and modulus ratio, rocks are designated as AH, BM etc.

2. *Rock Quality Designation (RQD)*

On the basis of core recovery, a quantitative expression of rock mass quality is proposed by Deere and coworkers (1967). The rock quality designation is the ratio of the cumulative length of the individual pieces exceeding 4 inches (10 cm) in length to the total length of the core expressed as percentage. The procedure for the estimation of RQD is indicated in Fig. 2.6.

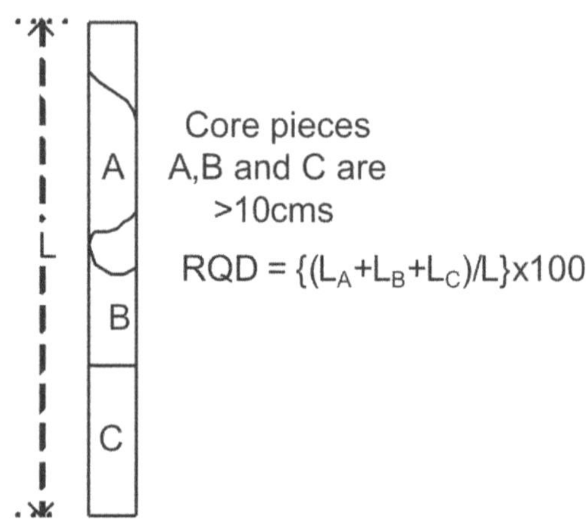

Fig. 2.6 Procedure for RQD estimation.

The rock formations are classified as :

	RQD (%)
Very poor	0-25
Poor	25-50
Fair	50-75
Good	75-90
Excellent	90-100

Deere and Deere (1988) have indicated the type of supports (rock bolts, steel ribs) which can be used for rock masses with RQD values ranging between 25 and 75. No supports are needed for formations with RQD values exceeding 75.

3. *Rock Structure Rating (RSR) :*

This has been proposed by Wickham and coworkers (1972) for a quantitative rating of the quality of a rock mass and also for the tunnel support recommendations. This classification system was the earliest to include shotcrete reinforcement in tunneling. The method incorporates three groups of parameters.

Group A : Geology
- Rock type (igneous / sedimentary /metamorphic)
- Rock hardness (hard / medium / soft / decomposed)
- Geological structure (massive / slightly folded / mostly faulted and folded / faulted / folded)

Group B : Discontinuities
- Joint spacing
- Joint orientation (strike and dip)
- Direction of tunnel drive as related to joint orientation

Group C: Groundwater inflow — As related to
- Overall rock mass quality on the basis of A+B together
- Joint condition (good / fair / poor)
- Amount of water inflow per unit time per unit length of the tunnel

In each group, for the parameters of the rock masses under investigation, ratings are assigned (Rating Tables of Wickham et al, 1972). The ratings, based on the type and nature of rock, for the geological structure varies from 0 (massive) to 8 (extremely deformed). Similarly in the case of group B, the ratings vary for joints as oriented to tunnel axis. If the strike of joints is parallel to tunnel, the ratings vary from 7 to 9. For joints with strike perpendicular to the tunnel, the ratings are from 9 to 13. For varying dips, the ratings are between 37 and 45. In the case of group C, the ratings for groundwater inflow are assigned depending on the sum of ratings of groups A and B on the following basis :

A + B	Heavy inflow	No inflow
13 - 44 :	6 - 12	12 - 22
44 – 75	10 – 18	18 – 25

The overall rating is estimated as the summation of values of all the groups (A+B+C). The final rating value thus obtained is utilized for the selection of the type and spacing of tunnel support requirements based on the graph presented by Wickham and coworkers.

This graph depicts different options in terms of steel ribs (spacing), rock bolts (spacing) and shotcrete (thickness) as related to various RSR values.

Table 2.3 Rock structure rating (RSR) parameters

Rock Structure Rating : Parameter A : General area geology

	Basic Rock Type				Geological Structure			
	Hard	Medium	Soft	Decomposed	Massive	Slightly Folded or Faulted	Modarately Folded or Faulted	Intensively Folded or Faulted
Igneous	1	2	3	4				
Matamorphic	1	2	3	4				
Sedimentary	2	3	4	4				
Type 1					30	22	15	9
Type 2					27	20	13	8
Type 3					24	18	12	7
Type 4					19	15	10	6

Rock Structure Rating : Parameter B : Joint pattern, direction of drive

	Strike ⊥ to Axis					Strike || to Axis		
	Direction of Drive					Direction of Drive		
	Both		With Dip	Agaira Dip		Either direction		
	Dip of Prominent joints[a]					Dip of Prominent Joints		
Average joint spacing	Flat	Dipping	Vertical	Dipping	Vertical	Flat	Dippng	Vertical
1. Very closely jointed, <2in	9	11	13	10	12	9	9	7
2. Clceety jointed. 2-6 in	13	16	19	15	17	14	14	11
3. Moderately pinted 6-12 in	23	24	28	19	22	23	23	19
4. Moderate to blocky. 1-2 ft	30	32	36	25	28	30	28	24
5. Blocky to massive. 2-4 ft	36	38	40	33	35	36	24	28
6. Massive, > 4ft	40	43	45	37	40	40	38	34

Rock Structure Rating : Parameter C : Groundwater, joint condition

Anticipated water inflow	Sum of Parameters A + B					
	13 - 44			45 - 75		
	Joint Condilion[b]					
gpm/1000 ft of tunnel	Good	F air	Poor	Good	F air	Poor
None	22	18	12	25	22	18
Slight, <200 gpm	19	15	9	23	19	14
Moderate, 200-1000 gpm	15	22	7	21	6	12
Heavy, >1000gp	10	8	6	18	14	10

[a]Dip : flat 0-20° dipping 20-50°; and vertical 50-90°

[b]Joint condition: good = tight or cemented; fair = slightly wethered or altered; poor = severely weathered, altered or open

4. *Rock Mass Rating System (RMRS)*

This has been proposed by Bieniawski (1976) and extended subsequently for use with guide lines for the choice of tunnel supports. The classification system involves the following six parameters:

1. Rock quality designation (RQD)
2. Unconfined compressive strength (UCS)
3. Joints - spacing
4. Joint conditions
5. Water saturation
6. Strike and dip of joints.

Table 2.4 Rock mass rating system (After Bieniawski, 1989) (Reproduced with permission)

A. Classification parametrs and their rating							
	Parameter		Range of values				
1.	Strength of infact rock material	Point-load strength index	> 10 MPa	4-10 MPa	2-4 MPa	1-2 MPa	For this low range-uniaxial compressive test is preferred
		Uniaxial comp. strength	> 250 MPa	100-250 MPa	50-100 MPa	25-50 MPa	5-25 MPa 1-5 MPa <1 MPa
		Rating	15	12	7	4	2 1 0
2.	Drill core Quality RQD		90%-100%	75%-90%	50%-75%	25%-50%	< 25%
	Rating		20	17	13	8	3
3.	Spacing of discontinuities		> 2 m	0.6-2 m	200-600mm	60-200mm	< 60 mm
	Rating		20	15	10	8	5
4.	Condition of discontinuities (see E)		Very rough surfaces not continuous no seperation unwealhered wall rock	Slightly rough surfaces seperation <1 mm slightly weal-hered walls	Slightly rough surfaces seperation <1 mm highly wealhered walls	Slickensided or Gauge <5 mm thick or seperation 1-5mm continuous	Soft gauge > 5 mm thick or Seperation > 5 mm continuous
	Rating		30	25	20	10	0
5.	Ground water	Inflow per 10m tunnel length (mm)	None	< 10	10-25	25-125	>125
		(Joint water press) / (Major principal σ)	0	< 0.1	0.1-0.2	0.2-0.5	> 0.5
		General conditions	Completely dry	Damp	Wet	Dripping	Flowing
		Rating	15	10	7	4	0

B. Rating adjustment for discontinuity orientations (See F)						
Strike and dip orientations		Very favourable	Favourable	Fair	Unfavourable	Very Unfavourable
Rating	Tunnels & mines	0	-2	-5	-10	-12
	Foundations	0	-2	-7	-15	-25
	Stopes	0	-5	-25	-50	

Table 2.4 *Contd....*

C. Rock mass classes determined from total ratings					
Parameter	Range of values				
Rating Class number Description	100 ← 81 I Very good rock	80 ← 61 II Good rock	60 ← 41 III Fair rock	40 ← 21 IV Poor rock	< 21 V Very poor rock
D. Meaning of rock classes					
Class number	I	II	III	IV	V
Average stand-up time	20 yrs for 15m span	1 year for 10m span	1 week for 5m span	10 hrs for 2.5m span	30 min for 1m span
Cohesion of rock mass (kPa)	>400	300-400	200-300	100-200	< 100
Fridion single of rock mass deg)	> 45	35-45	25-35	15-25	< 15
E. Guidelines for classification of discontinuity conditions					
Discontinuity length (persistence)	< 1 m	1-3 m	3-10 m	10-20 m	>20 m
Rating	6	4	2	1	0
Separation (aperture)	None	<0.1 mm	0.1-1.0 mm	1-5 mm	>5 mm
Rating	6	5	4	1	0
Roughness	Very rough	Rough	Sightly rough	Smooth	Sticknsided
Rating	6	5	3	1	0
Infilling (gauge)	None	Hard filling <5 mm	Hard filling <5 mm	Soft filling >5mm	Soft drilling <5 mm
Rating	6	4	2	2	0
Weathering	Unweathered	Slightly weathered	Moderately weathered	Highly weathered	Decomposed
Ratings	6	5	3	1	0

F. Effect of discontinuity strike and DIP orientation in tunnelling**			
Strike perpendicular to tunnel axis		Strike parallel to tunnel axis	
Drive with dip - DIP 45-90° Very favourable	Drive with dip-DIP 20-45° Favourable	DIP 45-90° Very favourable	DIP 20-45° Fair
Drive against dip - DIP 45-90° Fair	Drive against dip-DIP 20-45° Unfavourable	DIP 0-20 - Irrespective of strike Fair	

* Some conditions are mutually exclusive. For example, if infilling is present, the roughness of the surface will be overshadowed by the influence of the gauge. In such casees use A4 directly.

** Modified after Wickham et al (1972).

Each of these parameters for the rock mass is assigned a rating value, appropriately selected from the rating values in Table 2.4. For this purpose, the rock mass is divided into several regions usually bounded by structural features such as faults. All the rating values corresponding to the rock mass under consideration are noted from the respective data tables and added together to obtain the final value of the rock mass rating.

A set of guide lines, as proposed by Bieniawski (1989), for choosing the supports for a 10m – span tunnel are presented in Table 2.5. Appropriate to the RMR value obtained for the rock mass under consideration, the length, spacing and the type of supports can be chosen from these guidelines. However, these need to be further reinforced if new problems are encountered in the rock mass during the excavation.

Table 2.5 Guidelines for excavation and support of 10m span rock tunnels according to RMR system (Reproduced with permission) (After Bieniawski, 1989)

Rock mass class	Excavation	Rock bolts (20 mm diameter, fully grouted	Shotcrete	Steel sets
1-Very good rock *RMR : 81-80*	Full face, 3m advance			
II-Good rock *RMR : 61-80*	Full face,	Locally, bolts in crown 3m long, spaced 2.5 m with occasional wire mesh.	50 mm in crown where required.	None.
III-Fair rock	Top heading and bench 1.5-3 m advance in top heading. Commense support after each blast. Complete support 10m from face.	Systematic bolts 4m long, spaced 1.5-2m in crown and walls with wire mesh in crown.	50-100 mm in crown and 30 mm in sides	None.
IV-Poor rock	Top heading and bench 1.0-1.5m advance in top heading. Install support concurrently with excavation, 10m from face.	Systematic bolts 4-5 m long, spaced 1-1.5 m in crown and walls with wire mesh.	100-150 mm in crown and 100 mm in sides.	Light to medium ribs spaced 1.5 m where required.
V-Very poor rock *RMR : < 20*	Multiple drifts 0.5-1.5 m advance in top heading. Install support concurrently with excavation. Shotrete as soon as possible after blasting.	Systematic bolts 5-6 m long, spaced 1-1.5 m in crown and walls with wire mesh. Bolt inver.	150-200 mm in crown, 150 mm in sides, and 50 mm on face.	Medium to heavy ribs spaced 0.75 m with steel lagging and forepoiling if required. Close invert.

5. *Rock Quality Index System (Q- System) :*

Tunneling quality index has been proposed by Barton and coworkers (1974) for the classification of rock mass and for the tunnel support choice.

The factors involved in this method are:

1. Rock quality designation (RQD)
2. Joint sets number (J_n)
3. Joint roughness number (J_r)

4. Joint alteration number (J_a)
5. Joint water reduction factor (J_w)
6. Stress reduction factor (SRF)

The index 'Q' is estimated as

$$Q = [RQD/J_n] \times [J_r/J_a] \times [J_w/SRF]$$

The value of Q varies from .001 to 1000.

The rock mass characteristics are represented by the quotients in the above equation as :

RQD/J_n	:	Size of the jointed rock mass
J_r / J_a	:	Shear strength in between the blocks
J_w /SRF	:	Active stresses

Thus the index 'Q' is essentially a function of these three parameters.

Estimation of Q is carried out by substituting the appropriate values (selected from the rating tables), corresponding to the rock mass under consideration, for all the factors in the equation.

Table 2.6 Classification of individual parameters in tunnelling quality index (Q) (Reproduced with permission) (After Barton et al, 1974)

Description	Value	Notes
1. Rock auauty designation	RQD	
A. Very poor	0-25	1. Where RQD is reported or measured as <10 (including value of 10 is used to evaluate Q.
B. Poor	25-50	
C. Fair	50-45	
D. Good	75-90	2. RQD intervals of 5, i.e. 100, 95, 90 etc. are sufficiently accurate.
E. Excellent	90-100	
2. Joint set number	J_n	
A. Massive, no or few joints	0.5-1.0	
B. One joint set	2	
C. One joint set plus random	3	
D. Two joint sets	4	
E. Two joint sets plus random	6	
F. Three joint sets	9	1. For intersections use ($3.0 \times J_n$)
G. Three joint sets plus random	12	
H. Four or more joint sets, random, heavily jointed, 'sugar cube', etc.	15	2. For portals use ($2.0 \times J_n$)
J. Crushed rock, earthlikes	20	

Table 2.4 *Contd....*

Description	Value	Notes
3. Joint roughness number		
a. Rock wall contact		
b. Rock wall contact before 10 cm shear		
A. Discontinuous joints	4	
B. Rough and irregular, undulating	3	
C. Smooth undulating	2	
D. Sticken sided undulating	1.5	1. Add 1.0 if the mean spacing of the relevant
E. Rough or irregular, planar	1.5	greater than 3m.
F. Smooth, planar	1.0	
G. Sicken sided, planar	0.5	2. J_r - 0.5 can be used for planar, sicken sided
c. No rock wall contact when sheared		joints having lineations, provided that the
H. Zones containing clay minerals thick enough to prevent rock wall contact	1.0 (nominal)	lineations are oriented for minimum strength.
J. Sandy, gravely or crushed zone thick enough to prevent rock wall contact	1.0 (nominal)	
4. Joint alteration number	J_a	or degrees (approx.)
a. Rock wall contact		1. Values of x, the residual friction angle, are intended as an approximate guide to the mineralogical properties of the only alteration products, if present.
A. Tightly healed, hard, non-coltening impermeable filling	0.75	
B. Uanltered joint walls, surface standing	1.0	25-35
C. Slightly altered joint walls, non-softening mineral coatings, sandy particles, clay-tree disinterated rock, etc.	2.0	25-30
D. Silly-, or sandy-clay coatings, small clay-fraction (non-softening)	3.0	20-25
E. Sofetening or low-friction clay mineral coatings, i.e. kaolnite, mica. Also chlorite. talc, gypsum and graphite etc., and small quantities of swelling clays. (Discontinuous coatings, 1-2 mm or less)	4.0	8-10
5. Joint alteration number	J_a	or degrees (approx)
b. Rock wall contact before 10 cm shear		
F. Sandy particles, clay-tree, disintegrating rock etc.	4.0	25-30
G. Strongly over-consolidated, softening day mineral fillings (continuous <5 mm thick)	6.0	16-24
H. Medium or low over-consolidation, softening day mineral fillings (continuous <5 mm thick)	8.0	12-16

Table 2.4 *Contd....*

Description	Value	Notes
J. Swelling clay fillings, i.e. montollonite, (continuous <5 mm thick). Values of J_a depend on percent of swelling clay-size particles, and access to water.	8.0-12.0	6-12
c. No rock wall cotact when sheared		
K.. Zones or bands of disintegrated or crushed	6.0	
L Rock and clay (see G, H and J for clay	8.0	
M. Conditions	8.0-12.0	6-24
N. Zones or bands of silly- or sandy-clay, small clay fraction non-sofeting	5.0	
O. Thick continuous zones or bands of clay	10.0-13.0	
P. & R. (see G, H and J for clay conditions)	6.0-24.0	
6. Joint water reduction	H_W	approx. water pressure (kgf/cm^2)
A. Dry excavation or minor inflow i.e. <5 m locally	1.0	< 1.0
B. Medium inflow or pressure, occasional outwash of joint fillings	0.66	1.0-2.5
C. Large inflow or high pressure in competent rock with unfilled joints	0.5	2.5-10.0 1. Factors C to F are crude estimates; increase J_W, if drainage installed.
D. Large inflow or high pressure	0.33	2.5-10.0 2. Special problems caused by ice
E. Exceptionally high inflow or pressure at decaying with time	0.2-0.1	> 10 are not considered.
F. Exceptionally high inflow or pressure	0.1-0.05	> 10
7. Stress reduction factor		*SRF*
a. Weakness zones interesting excavation, which may cause loosening of rock mass when tunnel is exacavated		
A. Multiple occurrences of weakness zones containing day or chemically disintegrated rock, very loose surrounding rock any depth)		10.0 1. Reduce these values of SRF by 25-50% but only if the relevant shear zones influence do not interest the excavation.
B. Single weakness zones cotaining day, or chemically distegrated rock (excavation depth < 50 m)		5.0
C. Single weakness zones containing clay, or chemically distegrated rock (excavation depth < 50 m)		
D. Multiple shear zones in competent rock (clay tree), loose surrounding rock (any depth)		7.5
E. Single shear zone in competent rock (clay tree), (depth of excavation < 50m)		5.0
F. Single shear zone in competent rock (clay tree), (depth of excavation < 50m)		2.5
G. Loose open joints, heavily jointed or 'sugar cube', (any depth)		5.0

Table 2.4 *Contd....*

Description		Value		Notes	
8. Stress reduction factor				*SRF*	
b. Competent rock, rock stress problems					2. For strongly anisotropic virgin stress field (if measured): when $5 < \sigma_1/\sigma_3 < 10$, reduce σ_c to $0.8\,\sigma_c$ and σ_t to $0.8\,\sigma_t$. When $\sigma_1/\sigma_3 > 10$, reduce σ_c and σ_t to $0.8\,\sigma_t$, where σ_c - unconfined compressive strength, and σ_t - tensile strength (point load) and σ_1 and σ_3 are the major and minor principal stresses.
		σ_c/σ_1	$\sigma_1\sigma_1$		
H.	Low stress, near surface	> 200	> 13	2.5	
J.	Medium stress	200-10	13-0.66	1.0	
K.	High stress, very light structure (usually favourable to stability, may be unfavourable to wall stability)	10 - 5	0.66-0.33	0.5-2	3. Few case records available where depth or crown below surface is less than span width. Suggest *SRF* increase from 2.5 to 5 for such cases (see H).
L.	Mild rockburst (massive rock)	5-2.5	0.33-0.16	5-10	
M.	Heavy rockburst (massive rock)	< 2.5	< 0.16	10-20	
c.	Squeezing rock, plastic flow of incompetent rock under influence of high rock pressure				
N.	Mild squeezing rock pressure			5-10	
O.	Heavy squeezing rock pressure			10-20	
d.	Swelling rock, chemical swelling activity depending on presence of water				
P.	Mild swelling rock pressure			5-10	
R.	Heavy swelling rock pressure			10-15	

Additional notes on the use of these tables

When making estimates of the rock mass quality (Q), the following guidelines should be followed in addition to the notes listed in the tables :

1. When borehole core is unavailable, RQD can be estimated from the number of joints per unit volume, in which the number of joints per metre for each joint set are added. A simple relationship can be to convert this number to RQD for the case of clay free rock masses : RQD – 115 – 3.3 J_V (applax.), where J_V - total number of joints per m^3 ($0 < RQD < 100$ for $35 > J_V > 4.5$).

2. The parameter J_n representing the number of joint sets will often be affected by foliation, schistostly, staty cleavage or bedding etc. If strongly developed, these parallel 'joints' visible, or if only occasional breaks in the core are due to these leature, then it will be more appropriate to count them as 'random' joints whene valuating J_n.

3. The parameters J_r and J_a (repressing shear strength) should be relevant to the weakest significant joint set or day filled discontinuity in the given zone. However, if the joint set or discontinuity with in the given zone. However, if the joint set or discontinuity with the minimum valueo f J_r / J_a is favourably oriented or stability, then a second, less favourably oriented joint set or discontinuity may sometimes be more significant, and its higher value of J_r / J_a should be used when evaluating Q. The value of J_r / J_a should in fact relate to the surface most likely to allow failure to initate.

4. When a rock mass contains clay, the factor *SRF* appropriate to loosening loads should be evaluated. In such cases the strength of the infact rock is of little interest. However, when joining is minimal and clay is completely absent, the strength of the infact rock may become the weakest link, and the stability will then depend on the ratio rock-stress/ rock strength. A strongly anisotropic stress field is unfavourable for stability and is roughly accounted for as in note 2 in the table for stress reduction factor evaluation.

5. The compressive and tensile strengths (s_c and s_t) or the infact rock should be evaluated in the saturated condition if this is appropriate to the present and future in situ condition. A very conservative estimate of the strength should be made for those rocks that deteriorate when exposed to moist or saturated conditions.

Rock mass rating on the basis of Q value :

Quality	**Q value**
Excellent	400 – 1000
Extremely good	100 – 400
Very good	40 – 100
Good	10 - 40
Fair	4 – 10
Poor	1- 4
Very poor	0.1 – 1
Extremely poor	0.01 – 0.1
Exceptionally poor	0.001 – 0.01

The value of 'Q' is used for tunnel support requirements. For this purpose, a chart relating "Q" with another parameter designated as equivalent dimension (D_e) has been proposed by Barton and coworkers. Details are presented in chapter 9.

3

Weathering and Soil Formation

Weathering is a process involving disintegration and decomposition of the rocks in nature. The disintegrated and the altered products stay at the site of formation. If these are transported from the site with the help of natural agencies like wind and running water, the process is called erosion. Weathering and erosion go side by side involving removal of material and exposure of fresh surface for weathering again.

Weathering is categorized as a mechanical, chemical or biological type. In mechanical weathering, the process involves only fragmentation or physical breakdown of the rock into smaller fragments. On the other hand, chemical weathering involves chemical reactions resulting in the alteration of the rock leading ultimately to the formation of new alteration products. As fragmentation of rock material opens up new fresh particle surfaces (more surface area exposed for chemical alteration), the mechanical weathering promotes a faster rate of chemical weathering. Biological weathering is through organisms.

Mechanical Weathering

In nature, physical breaking of rocks is caused by several processes. All these involve widening of the fracture surfaces ultimately resulting in the detachment of the blocks surrounded by these weak planes. The processes operative are :

1. ***Frost-wedging :*** Water present in the cracks of the rocks freezes during the night time and thaws during day time. Freezing of water involves an increase in the volume (of about 9%) because of which the walls of the cracks are wedged (pushed apart).

2. ***Alternate expansion and contraction :*** In arid climates, there are drastic changes between the day and night temperatures. Surface heating results in thermal expansion during daytime and sudden drop in the temperature during the night time causes contraction. The temperature differential is around 25 to 30^0 C in deserts. It may be indicated that the expansion and contraction are confined only to the surface layers of the rock. This results often in the fracturing and detachment of the top layers. The process resulting in the removal of layers is also known as spheroidal weathering.

3. ***Fracturing through pressure release :*** Rocks at depth are confined under high confining uniform pressure with resistance to volume changes. However, if the rock material is uplifted due to tectonic processes to relatively lesser depths, it is subjected to lesser pressure conditions. This release of pressure brings in the release of internal stresses ultimately leading to the deformation of rock and generation of weak zones such as fractures. It is a common experience that in deep underground mining operations such as in the Kolar Gold Fields in Karnataka, release of the internal rock pressure during tunneling leads to sudden rock bursts.

4. ***Mineral crystallization from solutions in cracks :*** This is common in arid regions and similar to the frost action in terms of increase in the volume of the new product. When salt crystallizes from solutions hosted by the rock fractures, the increase in the volume of the salt minerals pushes or wedges the cracks. Thermal expansion of the salt minerals also introduces additional pressures within the cracks. The salt minerals also absorb water, which causes additional expansion in the volume. Thus the salt minerals undergo volume changes through different operations.

5. ***Effect of vegetation :*** During growth of vegetation in rocky terrains, the roots enlarge through the existing weak planes and in this process, with time, the cracks are widened often leading to the physical breakdown and loosening of the rock blocks.

Chemical weathering

As already indicated, this involves chemical decomposition and alteration of the rock with the formation of new products of weathering. Weathering of granite with the ultimate formation of kaolinite clay is a common example. Soil formation also involves chemical weathering. Water, CO_2, O_2 and organic acids are operative in the chemical decomposition and formation of new products.

The dominant processes active in chemical weathering are :

1. ***Hydration :*** This involves incorporation of water in the chemical composition of anhydrous mineral. Common example is the hydration of anhydrite ($CaSO_4$) to gypsum ($CaSO_4.2H_2O$). This reaction is accompanied by an increase in the volume up to about 60%. Formation of iron hydroxide (limonite) from hematite (Fe_2O_3) is another example, also resulting in an increase of volume.

2. ***Hydrolysis :*** Water dissociates into $(OH)^-$ and H^+ ions, which displace other ions in the mineral lattice, resulting in the formation of an altered product. Hydrolysis increases with the increase in temperature and with acidity in the environment. Hydrolysis is one of the processes involved in the formation of clay minerals. The $(OH)^-$ ions are bonded to Al^{3+} in an octahedral coordination. These Al-OH octahedral layers are in combination with the Si-O tetrahedral layers resulting in the formation of unit layers of clay minerals. The K^+ ions and silicic acid (H_4SiO_4) released during the reaction, are carried away in solution by the circulating waters.

3. ***Oxidation*** : The decomposition of minerals in a rock during chemical weathering is brought about by O_2 in the presence of water. An example of this type is the weathering of pyroxene to limonite through the following reaction :

$$4FeSiO_3 + O_2 + 2H_2O \rightarrow 4FeO(OH) + 4SiO_2$$

(pyroxene) (limonite) (silica in solution)

Oxidation of pyrite (FeS_2) is another example. Initially the iron sulphate mineral forms which ultimately converts into limonite. Alternatively, depending on the oxygen availability, pyrite converts into hematite (Fe_2O_3). The reactions are:

$$FeS_2 + H_2O + 7(O) \rightarrow FeSO_4 + H_2SO_4$$

$$2FeSO_4 + (O) + H_2SO_4 \rightarrow Fe_2(SO_4)_3 + H_2O$$

$$2FeS_2 + 15(O) + 4H_2O \rightarrow Fe_2O_3 + 4H_2SO_4$$

The sulphuric acid released promotes dissolution of minerals like calcite ($CaCO_3$) present in the weathered material. Oxidation of divalent to trivalent iron minerals is an exothermic reaction and also involves decrease in the density (increase of volume) of the product.

4. ***Carbonation*** : Surface waters contain CO_2 and its combination with water results in the formation of carbonic acid.

$$H_2O + CO_2 <-> H_2CO_3$$

$$H_2CO_3 <-> H^+ + HCO^{3-}$$

Production of carbonic acid lowers the pH, resulting in the attack of minerals.

Dissolution of rock-forming minerals like calcite takes place in the presence of H_2CO_3. The involved reaction is

$$CaCO_3 + H_2CO_3 \rightarrow Ca^{2+} + 2HCO^{3-}$$

This process is known as a congruent dissolution where the entire material goes into solution.

Conversion of feldspar to clay is another example. Weathering of feldspars involves the dissolution as well as the hydrolysis. Dissolution in this case takes place with the participation of selective components and is a case of incongruent dissolution. The reactions are of the type :

$$2KAlSi_3O_8 + 2H_2CO_3 + H_2O \rightarrow Al_2Si_2O_5(OH)_4 + 4SiO_2 + 2K^+ + 2HCO^{3-}$$

(orthoclase) (kaolinite) (dissolved silica)

$$2NaAlSi_3O_8 + 2H_2CO_3 + 9H_2O \rightarrow Al_2Si_2O_5(OH)_4 + 4H_4SiO_4 + 2Na^+ + 2HCO^{3-}$$

(kaolinite)

Biological weathering

Biological weathering involves breakdown of rock and also through chemical action by living organisms. Living organisms, besides burrowing action, also release organic acids. These include oxalic acid, phenolic acid, fulvic acid, acetic acid and humic acid. Bacteria and fungi are the most common microorganisms. Some of the microorganisms penetrate into mineral crystals and remove specific ions from the inter layers. Removal of K^+ from mica inter layers by fungi is an example of this type.

Susceptibility of Silicate Minerals to Weathering

Resistance of minerals to weathering depends upon their crystal structure. Susceptibility to weathering depends on the degree of shearing of Si-O tetrahedra in the lattice of silicates. Minerals formed at higher temperature during the early stages of magmatic crystallization are the most susceptible ones. The least susceptible are the ones at the last stage of crystallization. Thus, quartz is the most stable mineral.

The sequence of stability (from high to low stability in weathering), known as the Goldlich's stability series, is the reverse of the Bowen's reaction series. In terms of decreasing stability, the primary rock-forming silicates are ordered as follows :

Most stable ↓ Least stable

Quartz
Muscovite
K-Feldspar
Na-Plagioclase
Amphibole
Ca-Plagioclase
Pyroxene
Olivine

Olivine weathers rapidly because in its structure, the Si-O tetrahedra are weakly bonded to Mg-O linkages. In the chain and sheet silicate structures, the weakest centres are the cation-oxygen linkages. The sequence in this regard from weak to strong is Ca, Na and K respectively. Quartz is the most stable because it does not have any metal ion linkages and it has a three- dimensional Si -O tetrahedral structure.

Intensity of Weathering

The degree of weathering is controlled by several parameters. These are :

- Rock mass characteristics
- Climate
- Vegetative cover
- Relief
- Time
- Soil cover

Rock mass characteristics : These include composition of the rock, and its structural status. The ultrabasic and basic igneous rocks such as peridotite , dunite and gabbro with ferromagnesian minerals decompose rapidly in relation to acidic igneous rocks such as granite. The reasons for the same can be explained on the basis of Goldlich's stability series, detailed above. Similarly carbonate rocks weather rapidly due to the constituent mineral susceptibility to chemical solvents and dissolution. Slaking of shales in presence of water is common. Among the metamorphic rocks, quartzite is most stable. Weathering of schists and phyllites is relatively faster.

Structural characteristics in a rock mass play a crucial role in weathering. Jointed deformed rocks with folding and faulting undergo rapid weathering. The weak zones facilitate mechanical and chemical weathering by natural agencies.

Climate : Climate includes temperature and rainfall. In general, weathering is faster in regions with high temperature and high rainfall. Mechanical weathering is prevalent in arid regions with high temperatures. Further, as the temperature increases the vibration of atoms and ions in the rock mineral structure are more, ultimately leading to the development of cracks at the macro-level. Rate of chemical weathering doubles with an increase of temperature by 10^0 C. Rainfall contributes to the growth of organisms, bacteria and vegetation. Respiration of the living organisms produces CO_2 and vegetation contributes to O_2. Plant decay is common in humid regions.

Types of weathering as a function of temperature and pressure are indicated in Fig. 3.1 in terms of different climates.

Although the effects of temperature and pressure are detailed here separately, both these parameters operate in conjunction with each other in different climates. For example, in tropical climate with adequate rainfall and high temperatures (Fig. 3.1), weathering is rapid. Moist temperate climate ranks next in order. In the desert and tundra regions, the chemically altered material is scarce in general and the weathered layer is essentially of rock debris.

Weathering mantle in relation to climate is illustrated in Fig. 3.2.

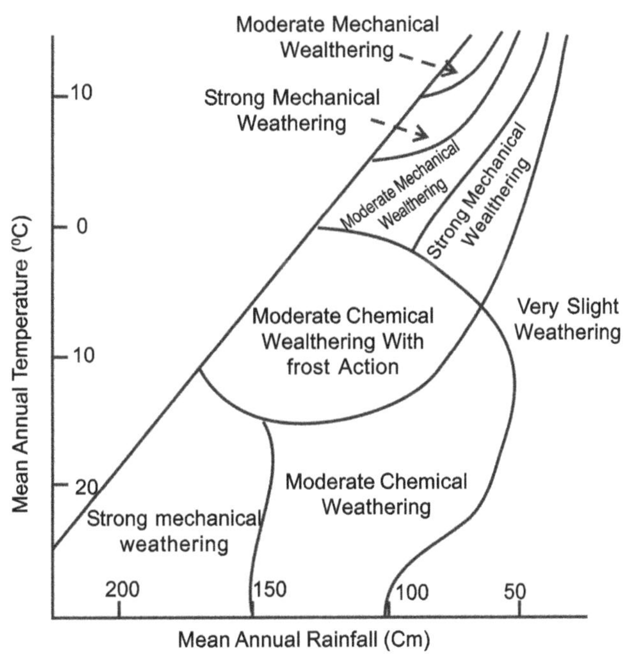

Fig. 3.1 Weathering as a function of temperature and rainfall (After Peltier, 1950)

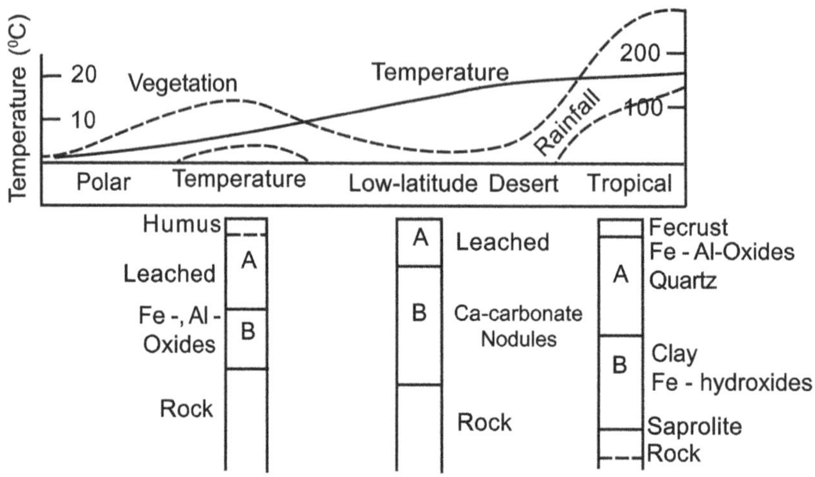

Fig. 3.2 Soil profiles at different latitudes (Modified after Strakhov, N.M., 1967)

Relief : If the topography is undulating and the slopes are steep, the weathered material, as it forms, is continuously eroded from the site. Consequently additional fresh surface is exposed to weathering. In addition, topography controls the rate of water percolation. Gentle slopes or flat surfaces retain water and facilitate availability of enough moisture for chemical weathering.

Time : If the weathering has continued over a long period of time, thick zone of alteration develops. For example, prolonged weathering results in the formation of thick bauxite deposit as a product overlying the alumino-silicate source rocks.

Soil cover : Soil itself is a product of weathering. But if a weathering continues in areas with soil cover, it further acts as an agent of weathering. Soil acts as a host to bacteria and organisms, which induce acidity in the water retained by the soil.

Laboratory Tests for Assessing Intensity of Weathering

To assess the degree of weathering, simple tests are adopted. These tests involve durability of the sample to wetting and drying (slake durability test) and also the extent to which the sample undergoes swelling in presence of water (swelling test).

Slake durability test : The rock material, in the form of pieces (of size around 5cm or more) is chosen. About 0.5 kg of the oven-dried and weighed sample is put in the rotary drum of the test equipment. The lower part of the drum is in water while it rotates through. The drum is rotated at a speed of 20 revolutions per minute for about 10 to 15 minutes. The outer wall of the drum has perforations of 2 mm mesh size to enable the material of size smaller than this, produced during the test to escape out of the drum. At the end of test, the material inside the drum is oven-dried and weighed once again. The durability is quantified in terms of the percentage of the material (by weight) retained in the test.

$$\text{Slake-durability index (\%)} = (W_1 / W) \times 100$$

where W = the original weight of the sample

W_1 = the weight of the sample after the test

Swelling pressure test : The sample in the form of a core specimen is subjected to water saturation and the change in its length is monitored through a set up (odeometer or equipment similar to it fitted with a sensitive dial gauge). The change in its length after its saturation for 12 hours is noted. The free swell coefficient (CS) is estimated as the ratio of change in length (ΔL) to the original length (L)

$$\text{Coefficient of swelling } (C_S) = \Delta L / L$$

The degree of swelling is finally indicated in terms of the swelling pressure.

Swelling stress (σ_S) = C_S. E where E is the deformation modulus

Engineering Classification of Weathered Rock Masses

Several procedures are proposed in the past based on the changes in the rock fabric, strength, extent of alteration along the discontinuities, free swelling, water absorption, extent of conversion of feldspars in alumino-silicate rocks on weathering, ultrasonic velocities of the rock masses and modal analyses of the mineralogical composition.

Broadly these classifications can be categorized into

- qualitative
- quantitative

Qualitative classifications

Among several of the methods proposed, the following two of the approaches are of importance.

Classification of weathered profiles (Deerman, 1974)

Weathering profiles are graded, from the top surface to the unaltered fresh rock at depth, into six grades (I to VI) with the highly altered top horizon designated as grade VI and the unaltered rock mass at the bottom as grade I. Changes in strength, permeability and deformability are also taken into consideration.

This approach is very useful for the preparation of weathering maps for project sites in Civil Engineering practice.

Weathering grade classification

This classification of the rock mass is proposed on the basis of the recommendations by the Engineering Geology Working Party of the Geological Society of London (1995).

The degree of weathering in a rock mass is arrived on the basis of the change in its strength, extent of alteration on its surface and also along the fractures or discontinuities. Alteration is noted in terms of discolouration or formation of altered products including soil.

In this simplified classification, the rock mass is categorized into six grades ranging from grade I (unaltered fresh rock) through grade VI (totally altered into residual soil).

Details are indicated in Table 3.1.

Table 3.1 Grades of weathering in a rock mass

Description	Grade	Basis of grading
Unweathered	(UW)	The rock mass is fresh . no change is noticed. There is no discolouration or alteration observed.
Slightly weathered (SW)	II	Not significantly weathered ; discolouration present only along cracks or discontinuities.
Moderately weathered (MW)	III	Rock mass altered significantly and is partly modified into soil. Discolouration of the material noticed along the cracks. Increase in the extent of fracturing and disintegration evidenced.
Highly weathered (HW)	IV	Material is discoloured and considerable loss of strength is observed. More than half of the material is converted to soil. Decomposition penetrates deeply inside the rock material.
Completely weathered (CW)	V	Original strength is considerably lost. The rock mass is changed to soil.
Residual weathered (RW)	VI	Original fabric destroyed on total conversion to soil.

Quantitative classifications

Of the many approaches for quantification of the weathering status, the geodurability classification proposed by Oliver (1979) is discussed in detail here. Essential aspects of few other methods are presented briefly in Table 3.2.

Table 3.2 Summary of selected quantitative classifications of weathered rock masses

Author(s)	Parameters	Scale
Lumb (1962)	Ratio of quartz to feldspar	$I_W = (R_S - R_0) / (1 - R_0)$ I_W : Weathering Index R_S : Ratio of quartz to feldspar in weathered material (soil) R_0 : Ratio of quartz to feldspar in original unaltered rock mass I_W : varies from 0 (fresh rock) to 1.0 for totally altered rock mass.
Iliev (1967)	Ultrasonic velocities	C (coefficient of weathering) $= (V_F - V_W)/V_F$ V_F = Velocity in fresh rock V_W = Velocity in weathered rock **Grade** — **(Coefficient)** — **Velocity (m/s)** Fresh — (0) — > 5000 Slightly weathered — (0-0.2) — 4000-5000 Moderately weathered — (0.2-0.4) — 3000-4000 Strongly weathered — (0.4-0.6) — 2000-3000 Very strongly weathered — (0.6-1.0) — <2000

Table 3.2 *Contd....*

Author(s)	Parameters	Scale			
Irfan and Dearman (1978)	Quick absorption (A) Bulk density (D) Strength (UCS)	- Quantitative Weathering Index for granite			
			A (%)	D (gm/m^3)	UCS (MPa)
		Fresh	<0.2	2.61	>250
		Partly stained	0.2-1.0	2.56-2.61	150-250
		Completely stained	1.0-2.0	2.51-2.56	100-150
		Moderately weathered	2.0-10.0	2.05-2,51	2.5-100
		Highly weathered	>10.0	<2.05	<2.5

Geo-durability classification

The rock mass classification proposed by Deere and Miller (1966) and modified by Bieniawski (1973), outlined in chapter 2, is further modified by Oliver by incorporating the swell parameter. For the intact rock material, uniaxial compressive strength (UCS) is plotted against free swelling index (change in length on swelling to the original length of the sample). The details are indicated in Fig. 3.3.

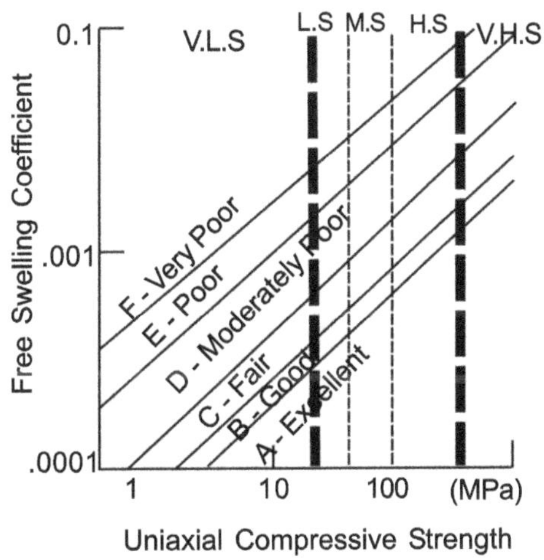

Fig. 3.3 Geo-durability Classification (Redrawn from Oliver, 1979)

(V.L.S: Very low strength, L.S: Low strength, M.S.: Medium Strength, H.S.: High strength V.H.S.: Very high strength)

This method is essentially meant for assessing the weathered status of the rock mass for tunnelling operations.

Soils

Soil is an important product of weathering. Broadly soils can be categorized as residual or transported types. The residual soil is formed in situ through weathering of parent rocks. It overlies the source rock at the site of its formation. Black cotton soil, formed from basalt, is an example of this type. On the other hand, transported soil is removed from its site of formation through erosion and transportation by natural agencies such as wind and water. Thus it has no relation with the underlying rocks at its place of occurrence. If the fluvial agency is involved, the soil is termed as alluvial soil. The alluvium in the Gangetic Plains is a classic example of this type. If the soil is transported and deposited by wind, it is known as eolian soil. Loess is a typical example of the same.

Several parameters contribute to the formation of soils by weathering. These include type of parent material, relief or topography, climate, time and biological activity. These are described in detail earlier (under intensity of weathering). Climate has a dominant role in the formation of different types of soil. In climates with adequate rainfall, the percolation of water to the subsurface promotes leaching of the soluble matter along with the colloidal material, which is transported to the next layer. Elluviation is the term used for the leaching and the process of deposition in the layer below is designated as illuviation. Soils formed under tropical climates with enough rainfall and temperature exhibit the distinct altered zones resulting from elluviation and illuviation. Distinct layers or zones are seen in a soil profile from the surface to the unaltered rock below at depth. The layers are called soil horizons.

Soil Horizons

These are layers in a soil profile which differ from each other in terms of texture, colour and mineralogical composition. The composition and the number of horizons depend on the climate, source material and topography. A typical soil profile with all the possible horizons is indicated in Fig 3.4.

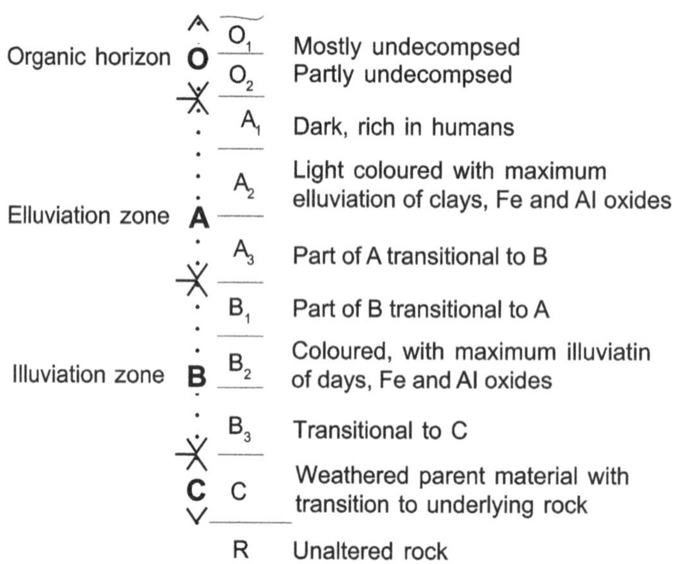

Fig 3.4. Soil horizons in a weathered profile

The organic matter contributes to humus and humic acid. The percolating water through the top organic zone consequently is acidic. This results in the leaching of soluble material and colloidal clay from the "A" zone (the zone of elluviation). This leached material is deposited in the zone of illuviation. Thus the zone of elluviation is porous compared to the zone of illuviation, which is relatively compact.

Classification of Soils

Soils are varied in their composition and properties depending upon the parameters involved and the process of formation. Several schemes of classification are available. Each of these classifications attempts to group the soils in terms of their genesis, properties or behaviour. Classification schemes help in ordering different soil groups and also catering to the requirements of the users. As already mentioned, soils are of interest to geologists, soil scientists, agricultural engineers and geotechnical engineers.

Classification schemes can broadly be categorized as

Natural classifications : based on genesis, attributes and profiles.

Technical classifications : based on properties and behaviour suited for engineering practice.

Natural Classification Schemes

These schemes deal with ordering of soil groups into a hierarchial system, into great groups, groups, subgroups, families and individual species. Among the existing natural classification schemes, some variation in ordering and the number of orders exist. But the general approach is similar. For use in engineering geology and geotechnical engineering, the natural classifications are not suitable because of the absence of inputs on engineering behavioural attributes. Hence a brief description of one such scheme is provided here to indicate the approach.

1. On the basis of processes involved, soils are broadly classified as

   - pedocal
   - pedalfer
   - saline soil
   - alkaline soil

   If the process involves precipitation and accumulation of calcium carbonate in the soil profile, the process is known as calcification. Such soils are acidic and known as pedocals (calcium-rich soils). These are typical of dry arid regions. The A-horizon is leached and calcium carbonate accumulates in the B-horizon. On the contrary, if calcium and other bases are removed through leaching, the process is known as podsolization. The resulting soil, called podsol, is a typical soil formed in humid climates. If intense leaching takes place under adequate rainfall and temperature (characteristic of tropical climate), iron and aluminum oxides are left after intense leaching of silica. This is typical of the lateritic soils and the process involved is known as lateritization. The laterites and podsols, being rich in aluminum and iron are known as pedalfers (iron and aluminum rich soils).

   Salinization is the process in which accumulation of Na, K, Ca and Mg salts takes place on the surface in dry climates with scarce rainfall, due to evaporation. This results in the formation of saline soils. If accumulation of sodium salts only occurs, the process is known as alkalinization. The soil formed through this process is known as an alkali soil.

The different soils formed under varying climatic conditions are grouped into zonal, intrazonal and azonal oders. Soils with distinct horizons, formed over extensive regions for a considerable period of time with adequate drainage are grouped under zonal order. The intrazonal oder covers soils formed under inadequate drainage conditions or in water-logged undrained conditions. Immature soils with no distinct characteristics are grouped under azonal order.

2. ***Geological classification based on mode of formation***

This classification is based on the mode of formation and natural agency active in weathering, erosion and transportation. The soils are classified into:

Soil type	Mode of formation	Grain size	Sorting
Fluviatile	Fluvial agency, river	Coarse – fine	Medium-good
Colluvial	Gravity, down the slope transport	Variable	Poor, Variable
Eolian	Wind	Medium –fine	Medium-good
Lacustrine	Lake	Medium-fine	Medium-good
Marine	Salt water, estuaries, lagoons	Medium-fine	Medium-good
Glacial	Ice, snow	Assorted sizes	Poor
Volcanic	Volcanic material	Coarse-fine	Poor-medium
Residual	In situ, weathering	Variable	Variable

3. ***Classification proposed by Food and Agricultural Organization (F.A.O.)***

This was first reported in 1974, on the basis of which the Soil Map of the World has been brought out by UNESCO in a simplified form. The scheme has been modified in 1988. Soil-forming parameters and diagonistic horizons are used for the classification. This is a widely used classification and has been modified to a certain extent by several countries to classify the local soils.

In the F.A.O soil classification scheme, based on the soil horizons present, physical and chemical properties as well as the colour, soils are categorized into orders, suborders, great groups, subgroups, families and series. In this 6-level hierarchy, ranking on the top, 10 orders are proposed. The World Soil Map prepared by FAO-UNESCO is based on these 10 orders. These 10 orders are subdivided into 47 suborders. In all 230 great groups are identified within these orders and suborders, of which 140 occur in tropical climate. Several hundreds of subgroups, families and series are proposed. This classification is used by soil and agricultural scientists.

Details of the soil orders are presented in Table 3.3.

Table 3.3 Soil classifications into soil orders

Order	Characteristics
Alfisols	with thin A-horizon, clayey B-horizon, with exchangeable cations (K, Na, Ca, Mg) greater than 50%.
Ultisols	with clayey B-horizon, Al-rich, extensively weathered, acidic with exchangeable cations less than 50% , formed under tropical and subtropical conditions
Oxisols	extensively weathered, highly oxidized, with considerable quantity of Fe and Al oxides, tropical rain forest environment
Vertisols	dark, with expansive clay, swelling soils, under tropical conditions
Mollisols	dark surface horizon (organic), with Ca exchangeable cation, typically prairie soil in grass-covered regions
Inceptisols	Well developed A-horizon with a weak B-horizon, young soils, example :volcanic soils with ash.
Entisols	Enough elluviation in A-horizon, light-coloured top layer, dissolved salts at depth.
Aridisols	Formed under arid conditions, with little or no organic matter, presence of caliche layer, typically desert and saline soils with very little plant growth
Spodosols	Formed under humid conditions in temperate regions, top horizons stained by organic matter
Histosols	Wet, organic-rich. Typical example: bog or muck

Technical Classifications

Classification Systems (ASTM, 1985; Unified Soil Classification System)

Soils are divided into coarse-grained and fine-grained types. In the former, more than 50% of the material is coarser than 0.075 mm size (i.e. retained on the #200 mesh) while in the latter, more than 50% is finer than this size (i.e. passes through the #200 mesh).

Atterberg limits (liquid limit and plastic limit) are used in the classification. Plasticity index (PI) is the difference between liquid limit (LL) and plastic limit (PL). The plasticity chart, originally proposed by Casagrande, has formed the basis for subsequent modifications. The classification on this chart is presented in Fig. 3.5.

Silt and clay are included in the fine-grained cateogory. Various symbols used are as follows:

Gravel	:	G
Sand	:	S
Silt	:	M
Clay	:	C
Organic	:	O
Peat	:	Pt
Well graded	:	W
Poorly graded	:	P
High liquid limit (>50)	:	H
Low liquid limit (<50)	:	L

Dual symbols are used for soil types in the border between two main types. The ML-CL symbol (Fig. 3.5), for example denotes silty clay.

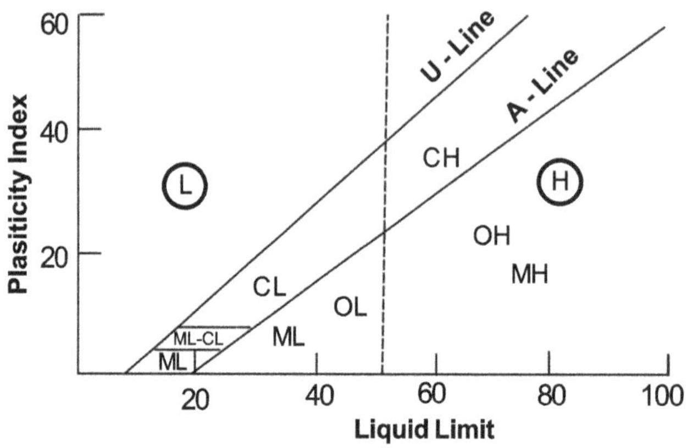

ML :	Sit		MH :	High Plasticity Silt
CL :	Clay		CH :	High Plasticity Clay
OL :	Organic Silt/Clay		OH :	Organic Clay/Silt

Fig. 3.5 Plasiticity chart for classification of soils.

For well-graded soils :

Gravel : $1 < C_c < 3$ and $C_u > 4$

Sand : $1 < C_c < 3$ and $C_u > 6$

The prefix "O" is used in the case of organic soils. The basis for the categorization of a soil as organic type is the liquid limit variation for the oven-dried sample as compared to the natural sample without oven drying (the oven-dried sample should have LL <75% of the LL of the natural one). As indicated in the figure , the U-line forms the upper bound for soils in general. The A-line demarcates the silty soils from the clay types.

Indian Soil Groups

Soils in India can be grouped under the following categories:

Group		**Corresponding taxonomic terms**
Alluvial soils	:	Entisol , Inceptisol , Alfisol
Lateritic soils	:	Alfisol , Ultisol , Oxisol
Red soils	:	Alfisol , Inceptisol , Ultisol
Black soils	:	Vertisol , Inceptisol , Entisol
Desert soils	:	Entisol , Aridisol
Salt-affected soils		
Forest soils	:	Alfisol
Peaty and marshy soils	:	Histsol

Alluvial soils : These cover approximately 24% of the land surface. The Gangetic plains region is an extensive alluvial tract. The alluvium composing these plains is a transported soil. The Ganga master stream has several big tributaries joining it from both the north and south directions. The material for the soil is derived from the rock formation on the foothills of Himalayas on the north and the hard rock terrains on the south. The alluvium is composed of sand and gravel interlayered with clay. The thickness of the alluvium in the central plains is enormous, of the order of several hundreds of meters. At places, local concentration of calcium carbonate concretions (called as kanker) occurs as a result of deposition of soluble carbonate material around sand grains acting as nuclei for the same. Two distinct types of alluvium are recognized. The older type (designated as Bhangar) is relatively dark in colour with significant amount of concretions (kankar). The clay content is also more. In contrast the younger alluvium (known as khadar) is light-coloured with more of granular material and less of clay.

Alluvial soils also occur in Narmada and Tapti valleys of Madhya Pradesh, Godavari and Krishna deltas in Andhra Pradesh. In Rajasthan, they occur in Alwar, Bharatpur , Sawai Madhopur and Ganganagar districts.

Alluvial derived soils occur along the coastal tracts on the east and west.

Lateritic soils : These are typical tropical soils derived from Deccan Trap basalts and basic igneous rocks. The texture varies from loam to clay. The top surface is highly porous (vesicular) and the soils are well drained. Often nodules of iron oxides and hydroxides are present. Due to extensive leaching, silica is removed and enrichment of iron and aluminum oxides and hydroxides takes place. A typical laterite profile is indicated in Fig. 3.6.

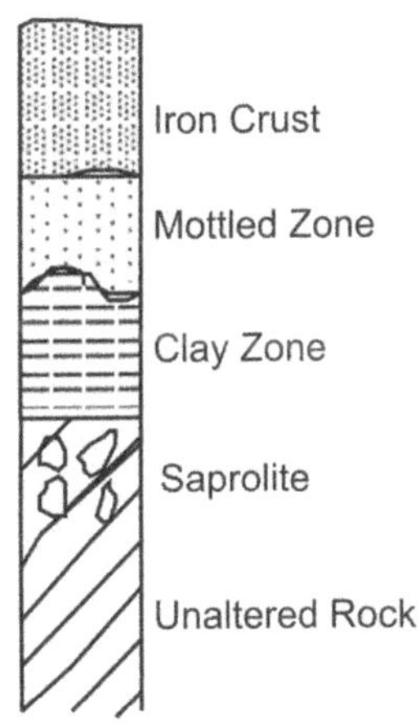

Fig. 3.6 A typical laterite soil profile.

Laterite soils occupy an extensive area in Maharashtra as also in parts of eastern Madhya Pradesh, Bihar, Orissa, West Bengal, Karnataka and Tamil Nadu. In Kerala, the occurrences are in Palakkad, Ernakulam and Alappuzha districts. Blocks of lateritic material are trimmed into shapes and used for construction purposes. In Assam, laterite occurs in large portions of Khasi and Jaintia hills.

Red soils : These soils are formed by residual weathering of rocks rich in ferromagnesian minerals. The soil is loamy to sandy with localized occurrence of carbonate concretions. These soils occur in Tamil Nadu (in Dharmapuri, Salem, South Arcot, North Arcot, Chingelput, Anna, Madurai, Pudukkotai, Kottabomman and Kanyakumari districts), Andhra Pradesh (Rayalaseema region), Rajasthan (in Sawai Madhopur, Udaipur and Dungarpur districts), Orissa, Madhya Pradesh and Assam (entire Mizo hills, parts of Cachar, Garo, Khasi and Jaintia hills).

Black soils : These are typical residual soils derived from basalts and basic igneous rocks. Popularly known as black cotton soil (also as regur), the soil is dark in colour, fine-grained, clayey in composition and rich in Mg- and Al- oxides. Significant amounts of calcium and magnesium carbonates occur in the soil profile. Organic material is absent.

The soil has montomorillonite clay, which is responsible for its expansive nature. The soil is problematic as it swells when wet and shrinks drastically on drying. It has high plasticity. Because of poor draining, often waterlogging is evidenced in the terrain.

The black soils occur extensively in Maharashtra extending to western Madhya Pradesh and parts of Gujarat. In Andhra Pradesh, occurrences are reported in Telengana and Rayalaseema regions. Often calcareous concretions are evidenced and the soils are saline at places.

Black soils mixed with red soils occur in easter Rajasthan (in Udaipur, Chittorgarh , Dungarpur, Banswara and Bhilwara districts).

Desert soils : These soils, of pedocal type, have calcium carbonate accumulation in the form of pellets and nodules in the profile in B and C horizons. They are devoid of organic matter. The soils are sandy with around 5% clay and are brownish in colour. Presence of nitrates and phosphates is common. In Rajasthan these soils occur in Jaisalmer and Bikaner districts.

Sand dunes, typical Aeolian deposits, cover substantial portions of Jodhpur, Barmer, Jaisalmer and Bikaner districts in Rajasthan.

The cold desert regions in Himalayan terrain at high altitudes have typical dark-coloured soils. These soils exhibit typical weathered landforms (Fig. 3.7). The high altitude desert regions of Ladakh in Jammu-Kashmir and Lahaul, Spiti, Kinnaur and Brahmour regions of Himachal Pradesh have these soils. These are typical rain-shadow areas.

Fig. 3.7 Typical weathering in cold desert areas of Spiti Valley, H.P. (Photo : Sita).

Salt-impregnated Soils

Saline soils and alkaline soils belong to this category. The saline soils are dominated by soluble salts like NaCl, Na_2SO_4, $CaCl_2$, $MgCl_2$, $CaSO_4$ and $MgSO_4$. The pH of the saline soil is less than 8.2. On the other hand, the alkaline soils mainly contain Na_2CO_3. Salts with other cations are absent. The pH of the alkaline soil is higher than 8.2.

The sources for the salt are diverse :

1. Ineffective leaching of the accumulated salt during weathering under arid and semi-arid conditions.
2. Waterlogging in canal-command areas and saline enrichment close to the surface due to constant evaporation of the subsurface water.
3. Excess of irrigation and fertilizer input without proper drainage and lack of circulation in the subsurface.
4. Ingress of saline water due to tidal invasion on to the delataic and inland areas.

These soils occur in parts of Gangetic Plains in U.P. , Punjab and Haryana as also in some water-logged canal-command areas in Andhra Pradesh. In Jammu and Kasmir, these soils occur in Leh, Pulwara and Patgam districts. Deltaic regions in Sunderbans (West Bengal) and Mahanadi (Orissa) have saline soils.

Forest soils :

These soils are dark brown in colour and rich in organic matter. The subsoil humus present results in the acidic nature of these soils. These occur in Terai region of Uttaranchal, parts of Punjab, Himachal Pradesh, Jammu-Kashmir and Tamil Nadu.

Peaty and Marshy soils :

These soils are rich in organic matter and acidic (with pH around 3.5). They are dark in colour. Because of the presence of ferrous iron, the soils exhibit bluish tint. They are clayey with free alumina and ferrous sulphate. The soils occur in coastal areas of Orissa, S.E. Tamil Nadu coast and in the Sunderbans delta (West Bengal).

Peaty soils occur in Kerala (Kottayam district).

4

Geomorphic Processes and Landforms

On the surface of the Earth, several geological processes are continuously at work. These are responsible for the evolution of landforms. These landforms have distinct morphology (forms) and the constituent material possesses specific characteristic properties. The subject which deals with these processes and evolution of landforms is known as geomorphology. The surface geological agencies that are at work include running water (fluvial), snow (glacial), sea (marine), wind (eolian) and arid climate (desert).

Fluvial Processes

In the hydrologic cycle, surface water is an important component. The surface flow is termed as the runoff in the form of streams. Stream is a general term used for all the surface flows including rivers and channels of different sizes. Streams are of two categories. They are either perennial or ephemeral (intermittent) depending on the flow. If there is enough flow althrough the year, it is known as a perennial stream. The Ganga, Brahmaputra and Indus are perennial in that the flow is from rainfall in the monsoon as well as from snow melt in the summer season. In fact, the rivers, which originate in the upper reaches of Himalayas are of this type. In contrast, the flow in the ephemeral or intermittent streams is contributed by the rainfall. Several of them even dry out or have very lean flow in summer. The rivers in Peninsular India are of this category.

Stream Flow

The flow in a stream controlled by three factors :

1. The stream gradient
2. Discharge or the quantum of water
3. Shape and roughness of the channel

The flow velocity increases with the gradient and discharge. Channel roughness and the shape result in the friction or resistance to the flow. Flow is more in a smooth channel compared to a rough one. Among the different channels with equal cross-sectional areas, the semi-circular channel has less contact surface with water and hence offers relatively less frictional resistance to the flow. The velocity is maximum in the central part than at the sides of the channel. The velocity profile in a stream is indicated in Fig. 4.1.

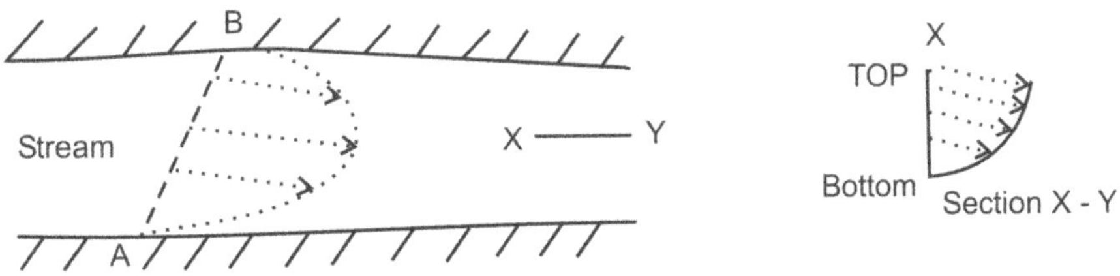

Fig. 4.1 Velocity profiles in a stream

Stream flow is either of a laminar or a turbulent type (Fig. 4.2). In a laminar flow, the stream path lines are straight and parallel to one another without crossing each other. In a turbulent flow, the stream lines mix and the flow has eddies.

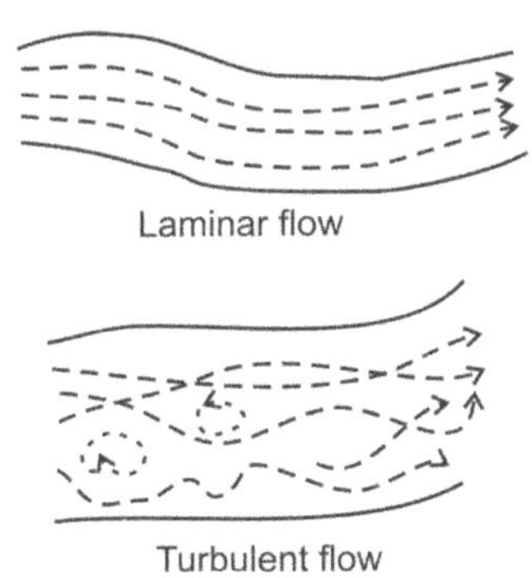

Fig. 4.2 Laminar and turbulent flows.

The turbulence flow is facilitated by an increase in the velocity of the stream. In addition, the depth of the channel and the viscosity of the stream water (offering resistance to the flow) also promote turbulence.

In stream hydraulics, Reynold's number and Froude number are used for expressing the flow type.

R_e (Reynold's number) = $\rho_w . v. r / \mu$,

where ρ_w = density of water

v = velocity, μ = dynamic viscosity

r = hydraulic radius of the channel

= A / L_p

A = area, Lp = wetted perimeter)

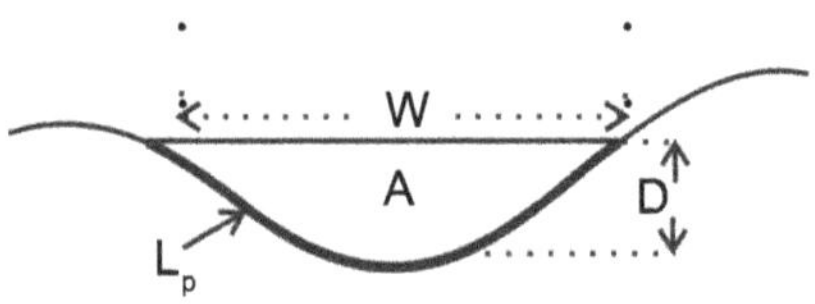

Channel section

Fig. 4.3 Channel parameters

For laminar flow, the value of Reynold's number should be less than 500.

F_r (Froude's number) = $v / \sqrt{(gD)}$, where D = channel depth,

g = acceleration due to gravity

For a subcritical flow, the value of F_r is less than 1. If it exceeds 1, the flow is supercritical. In nature, the flow is rarely supercritical. Only high velocity streams have supercritical flows.

The velocity (v) is related to the quantum of flow (Q).

Q = A. v where A = area of cross-section of the channel

$v = c.\sqrt{(r.s)}$ where r = hydraulic radius, s = slope and c = constant

The velocity, v, is the mean velocity at 0.6 of the depth from the surface of the stream.

Drainage Basin

This is defined as the area within which the main master stream has its contribution of water through overland flow and through groundwater. The overland flow is through the tributaries and streams that join these tributaries. Thus the drainage basin is the watershed region for the main stream. Each watershed or a drainage basin is demarcated by a drainage divide, encompassing all the streams that contribute to the master stream (Fig. 4.4). Beyond the drainage divide, the streams that occur contribute to some other master stream.

Within a drainage basin, streams have a hierarchial order ranging from the single small stream to a tributary with several streams joining it. Thus the entire network of streams is divided into several segments, which are defined by stream junctions that are designated by an integer number indicating the order of magnitude. Every single stream from its origin to its first confluence with another such stream is designated as the first order stream. When two first order channels meet, they generate a second order stream from that junction onwards. Similarly a third order stream starts from the junction of two-second order streams and so on. This stream hierarchy enables the analysis of a drainage basin.

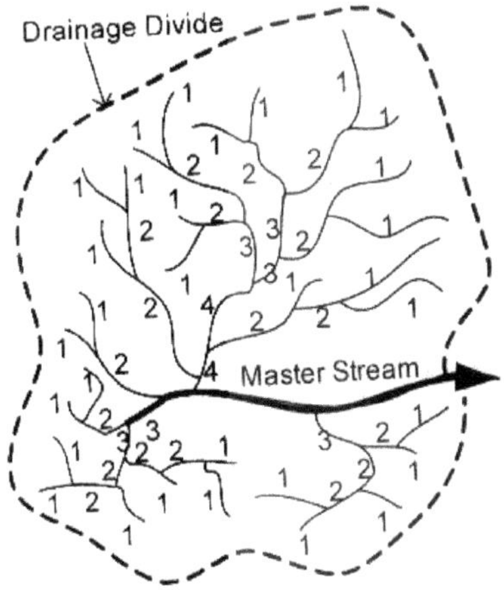

Fig. 4.4 Drainage basin of a master stream (numbers indicate stream orders).

Drainage Basin Analysis

The following parameters are used to analyze the drainage network in a drainage basin.

Bifurcation ratio : This is the ratio of number of streams of a given order to the number of streams of the next higher order. This ratio normally is between 3 and 5.

$R = (\Sigma N_n) / (\Sigma N_{n+1})$, where R = bifurcation ratio,

N_n = number of streams of order "n"

N_{n+1} = number of streams of order "n+1"

Length ratio : this is expressed as the ratio of total length of streams of any given order to the total length of all the streams of the next lower order.

$Lr = (\Sigma l_n) / (\Sigma l_{n-1})$, where l_n = length of streams of order "n"

l_{n-1} = length of streams of order "n-1"

Drainage density : This is the ratio of total length of all the streams (of all orders) within the basin to the area of the basin. Drainage density (D) is controlled by the type of formations in the basin. Areas with impervious formations will have higher drainage density than those with pervious formations.

$D = (\Sigma l) / A$, where Σl = total length of streams of all orders

A = area of the watershed or the drainage basin

Drainage Patterns

The spread of the channel network in a terrain is controlled by the topography, lithology of the formations and the geological structure. While the topography guides the channel path based on the relief and the slope directions, the lithology of the formations helps in the distribution (spread) of the network. As already explained, the drainage bifurcation is wide spread in the areas with impervious formations in contrast to the areas with permeable formations. Further, the competent formations offer more resistance to the channel cutting. Weak formations offer scope for the channel development. Structural control of the drainage network results in oriented and preferential spread of the drainage network along weak zones such as lineaments, joints and faults. For example, the Narmada master stream flows along a major lineament (linear weak zone) commonly known as the Narmada lineament.

Drainage patterns are broadly classified into six types. Details are presented in Table 4.1.

Table 4.1 Drainage patterns

Type	Controlling features	Recognition and characteristics	Typical pattern
1. Dendritic	Generally developed over massive impervious igneous and metamorphic rocks; characterized by lack of preferred orientation and spacing.	Resembles complex branching of a tree	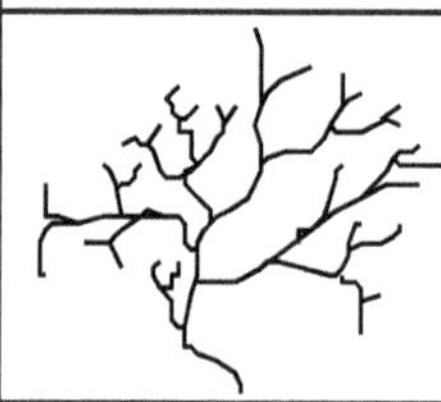
2. Trellis	Controlled by the structure of the formations; normally along the strike direction in folded rocks	Major parallel streams with short tributaries resembling short branches of a vine	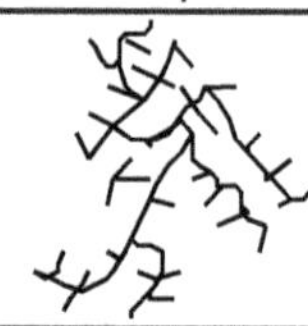
3. Radial	Controlled by the topography; the streams flow outward (centrifugal drainage) in areas of dome topography; the flow is radial to the centre (centripetal drainage) in the basin areas	Resembles the spokes of a wheel	
4. Parallel	Developed in rocks with parallel faults or weak zones; major stream occupies a fault while the tributaries are parallel and meet the master stream at approximately the same angle	Resembles a horse tail	
5. Annular	Similar to radial type, but in addition the streams occupy the arcuate or circular boundaries of different strata with domal structures	Ring-like tributaries that intercept the radial streams	
6. Rectangular	Controlled by joints; main channels with rectangular bends; tributaries join at right angles	Rectangular bends	

Stream Erosion and Sediment Transport

Stream energy plays the crucial role in erosion and sediment transport. This energy is proportional to the discharge (quantum of flow) and the flow velocity. The competence of the stream is indicated by the load it carries. For example, the load carried by the river Ganga in its upper reaches has a large proportion of very coarse material including boulders and pebbles, while in the Gangetic Plains it is essentially of sand, silt and clay fractions. In a deep stream, velocity controls the competence while discharge plays a significant role in the case of a shallow stream.

Stream erosion involves three operations

1. ***Hydraulic action :*** In this case, the erosion of the sediment is by the flowing stream itself. The flowing water dislodges material. It is common experience in times of flood that the fast moving floodwater erodes the banks of the channel causing considerable devastation.
2. ***Abrasion :*** This process involves the grinding of particles against each other and also against the coarse rock formations on the channel bottom. The sizes of the particles get reduced due to abrasion. The particles are also rounded, losing all the angularities. On the channel bottom, potholes form due to scouring action.
3. ***Solution :*** Much of the soluble material gets dissolved even from the weathering of the parent material itself. During the stream flow, the dissolved material is transported as solution load.

Thus the load carried by a stream is of three types:

1. ***Bed load :*** The coarse material like the pebbles are moved along the channel bottom by rolling or traction. The stream competence in such a case is that it can push these coarse particles but cannot lift them. Such a load is known as the traction load or bed load. In addition, if the particles are lifted to new sites in the vicinity due to sufficient stream velocity, the process is termed as saltation (Fig. 4.5).

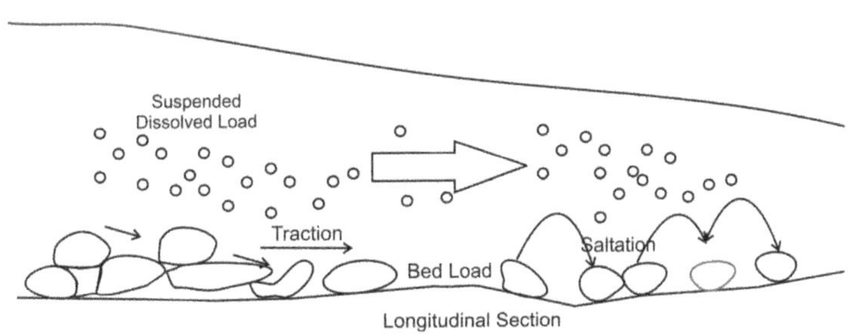

Fig. 4.5 Types of sediment load in a stream.

2. ***Suspended load :*** This load generally consists of sand, silt and clay sizes. In a flood, the quantum of the suspended load and its size ranges vary. Deposition of the suspended load is controlled by the mechanical and physical parameters such as the stream velocity and the gradient of the channel. If the velocity is reduced, part of the suspended load is deposited depending upon the competence of the stream.

3. ***Dissolved load (in solution):*** The deposition of this material is controlled by the chemical parameters in the basin. For example, in the geological past, climates varied from time to time. This has resulted in the selective precipitation or retention of the material of the solution load. To cite a typical example, the banded iron formations (the source material for iron ores) are composed of alternating bands of iron oxide (hematite or magnetite) and silica (quartz). The layers vary in their thickness. These are chemical sediments. Iron oxide precipitates in an alkaline environment while the silica precipitation needs acidic environment. Thus the alternate layers of iron oxide and silica in these formations indicate alternation of alkaline and acidic conditions in the environmental framework of their deposition. The thickness of each of such layers denotes the duration or prevalence of the relevant climatic condition.

As already explained, the current velocity is the controlling parameter for erosion, transportation and deposition (of the mechanical load and not the load in solution) in a fluvial environment. This is well represented in a graph, known as the Hjulstrom diagram (Fig.4.6).

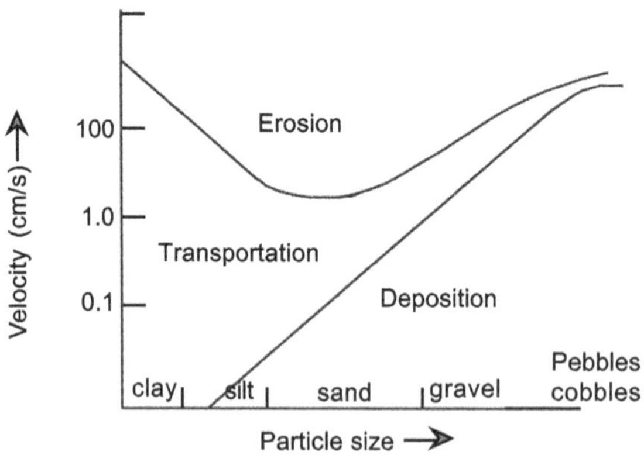

Fig. 4.6 Stream processes as related to velocity (Hjulstrom curve).

In the figure, it can be noticed that for erosion of clay and fine-silt sizes, higher velocity is needed. This is possibly due to the resistance offered by the finer sizes due to their cohesion. Of all these size grades indicated along the x-axis, the particles of sand size range are the ones that are eroded at relatively low velocities.

Graded stream : The gradient of a stream constantly modifies depending upon the processes of erosion and deposition. As the material is removed, transported and deposited at different places, the gradient or the profile changes. The stream flow takes place because of its gradient and the ultimate destination is the sea.

The theoretical base line for the stream is thus the sea level. A base line is defined as the imaginary extension of the sea level. As long as the stream gradient is above the base line, the stream processes of erosion, transportation and deposition are operative in a fluvial system.

However, with time the profile gradient gets modified into a typical concave profile (Fig. 4.7), steeper upstream and nearly flat towards its confluence with the sea.

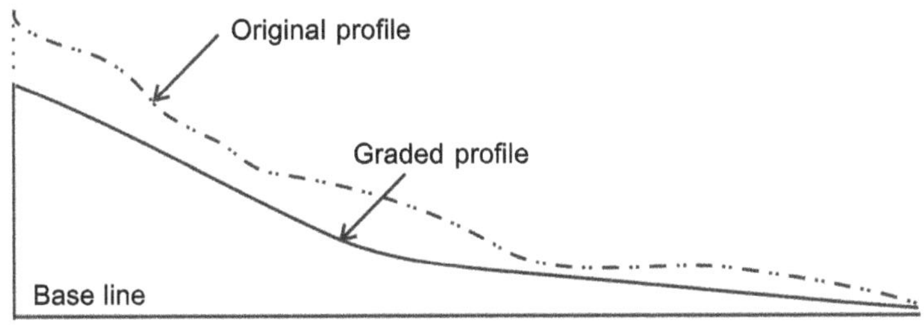

Fig. 4.7 Typical profile of a graded stream.

The stream profile remains almost unchanged if rate of channel erosion is equal to the rate of channel deposition. Thus a balance or equilibrium is attained. The stream under this condition has just the load that it can carry with least erosion or deposition. The stream with this equilibrium condition is called a graded stream. A graded stream operates under a dynamic equilibrium since any change in its operating parameters can alter the equilibrium. Then again the stream tends to attain its graded status over a period of time by constant modification to attain a graded profile.

Stream channel types

Channels can be categorized under three types based on the mode of formation and geometry. The controlling parameters are the slope, quantum of flow and the nature of material through which the channel is cut.

1. ***Straight channels :*** The channels are with sinuousity (ratio of stream length to the straight length across the horizontal stretch or the meander belt axis : Fig. 4.8) less than 1.5. They form on low slopes. In general, straight channels are not very common. The channel formation and development are guided by the nature of rock formations, geological structure, topography and the stream characteristics.

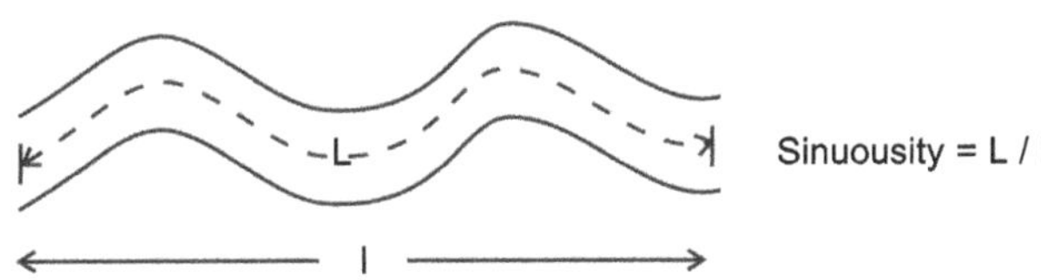

Fig. 4.8 Sinuousity of a meander loop.

2. ***Meandering channels :*** These are common in broad flood plains with fine-grained unconsolidated materials. The channel is characterized by several bends or meanders. Channels with sinuousity greater than 1.5 are classified as meandering channels. The stream cuts or erodes at the outer bend (Fig. 4.9) and deposits material towards the inside bend. Accordingly the channel geometry varies. The channel at the bends is asymmetric with the deeper portion of the channel located towards the outer bend.

Midway between the bends, the channel is symmetric and is designated as a riffle. The line indicating the deepest portion along the channel is known as thalweg (Fig. 4.9).

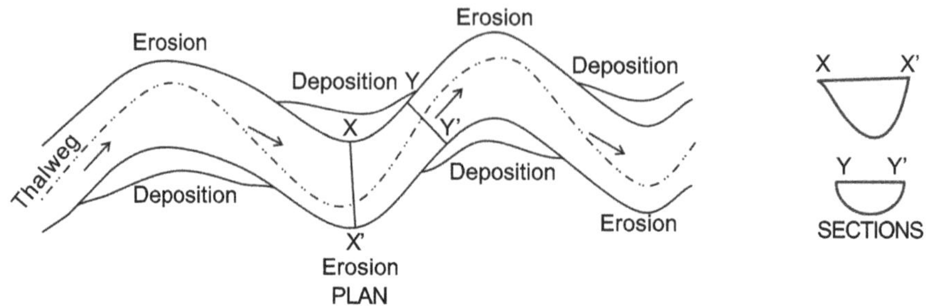

Fig. 4.9 A typical meandering channel.

Meanders in areas of uplift have narrow and deep channels. These are known as entrenched meanders. The channel cutting is in progress as the area is undergoing uplift and hence the channel is cut to depth. Thus entrenched meanders are typical of areas with neotectonic uplift. The Yamuna channel in the terrain south of the main Ganga stream in U.P. has typical entrenched meanders.

3. ***Braided channels :*** These are wide, straight and shallow channels with the stream bifurcating into several branches going around the obstacles (Fig. 4.10). The channel is quite wide compared to its depth. The braided channels develop due to large influx of material, commonly at the confluence of tributaries with the master stream. Because of velocity changes, the load brought by the tributary gets deposited a little beyond the confluence point. For example, the channel of master stream Ganga is braided on the downstream side at the confluence of Yamuna (joining it from the South) at Allahabad. In between the braided channels the islands are developed.

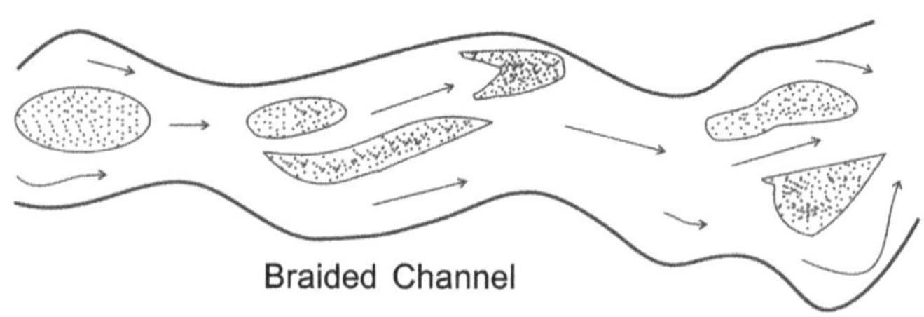

Fig. 4.10 Braided channel branching out around obstacles.

Cycles of erosion :

Different stages of evolution of drainage, valley development and landforms in a fluvial system are related to three distinct stages. The early stage is denoted as youth with the intermittent and final stages referred to as the maturity and old age. Each of these stages is characterized by distinct fluvial features.

Youth stage : The stream has high energy. It goes through rapids and falls. Development of V-shaped valleys with steep sides is common. The sediment load has a dominant coarse fraction.

Maturity stage : The valleys broaden into U-shapes and the flood plain develops. The drainages form into a well-integrated network. The rivers have meandering channels.

Old age : The valleys have gentle slopes. The relief is very less. The energy of the stream is highly reduced. A broad flood plain develops. River meandering in the broad flood plain is common. The sediment load has the sand, silt and clay fractions.

Fluvial Landforms

1. ***Flood plain :*** During the floods , a stream flows over its banks on to the adjacent land surface depositing the fine material. This occupied area is known as the flood plain. The term "peneplain" is used for a broad featureless flood plain, typical of the old age of the erosion cycle of a stream.

2. ***River terraces :*** These are the raised sections of the flood plain during repeated floods. They have step-like appearance. These represent old abandoned flood plains at a higher level. Thus terraces are indicative of flood levels. Terraces are of two types, namely unpaired (single) and paired terraces (Fig. 4.11).

   The unpaired terraces are formed by a meandering stream. As the meandering stream degrades into a flood plain, these terraces are preserved indicating its former flood plains. They are developed on one side of the stream.

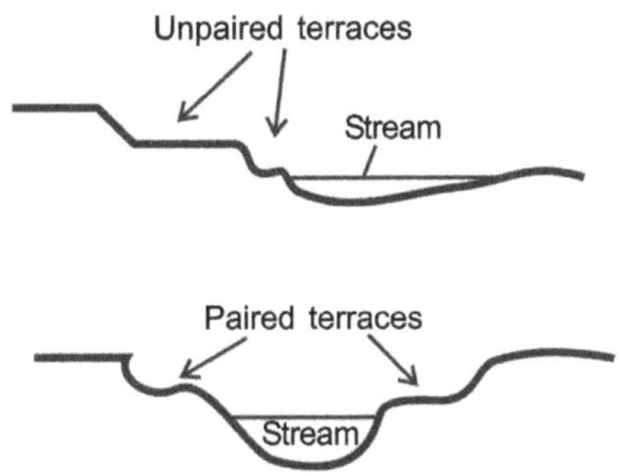

Fig. 4.11 Unpaired and paired terraces.

   Paired terraces are evidenced at the same elevation on either side of the channel (Fig. 4.11). These are formed by repeated down cutting by the stream with increase in its discharge.

3. ***Levees :*** These are raised surfaces on either side of the stream with gentle slope towards land side (Fig. 4.12). These are the low embankments formed from the deposition of sand and silt during flood. The floodwater overflows the banks of the channel laying the embankment of a low relief. Due to the confinement of the channel, the water level in the channel rises and during the subsequent flood, the overflow is quite substantial flooding the area beyond the levees.

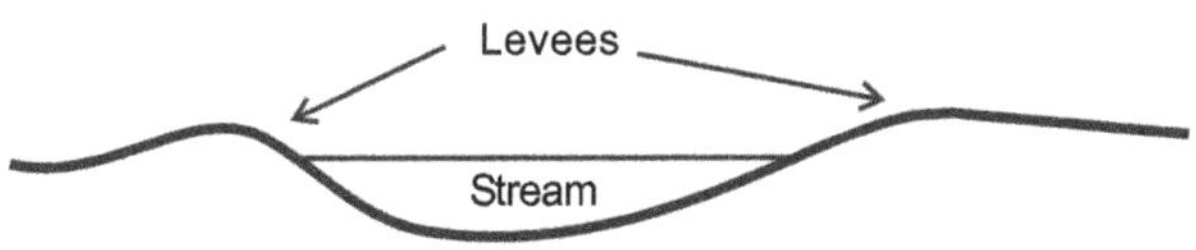

Fig. 4.12 Flood plain levees of a stream.

4. ***Oxbow lakes :*** These are the abandoned channels formed by meandering of a stream. As already explained earlier, a meandering stream has flow through channel bends. At the outer bends as the erosion progresses, the cutting at the loops from both sides results in opening a shorter path to the flow bypassing the loop (Fig. 4.13). Once the new channel is opened, the loop gets disconnected with deposition of finer material. The old abandoned meander loop forms a lake. This lake is known as an oxbow lake. The process of formation of an oxbow lake is indicated in Fig. 4.13. Oxbow lakes are clearly seen on the satellite imageries. Oxbow lakes are indicative of downstream migration of a channel.

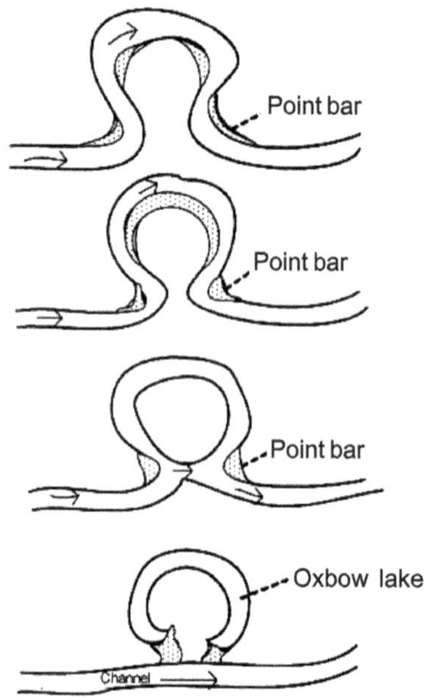

Fig. 4.13 Formation of an oxbow lake.

5. ***Point bars :*** These are the deposits formed towards the inner banks of a meandering channel (Fig. 4.13). The point bar slopes towards the thalweg. Thalweg, as indicated earlier in the Fig. 4.9, is a line joining the deepest portion of the channel. Point bars are composed mostly of sand and fine silt with minor amount of clay. Because of their gentle slope, the deposits often exhibit cross bedding. A gradation of size from coarse at the bottom to finer at the top is quite common.

6. ***Braided bar deposits :*** These are associated with the braided channels (Fig. 4.10). As explained earlier, the braided channels are due to the branching of the main channel to bypass the accumulated sediment load at places with its path. Because of their multi-channel nature, there is enough channel-fill material. They often exhibit convex geometry with curved surfaces. Braid bar deposits occur as single bars or as alternate bars (mid-channel bars). Alternate channels may join at places to form braid bars.

   Braided bar deposits grow laterally and also to downstream direction with modification of the channel directions. On satellite imagery, the directions of growth are clearly evidenced from the vegetational contrast. For example, in the Ganga masterstream,on the downstream side of the confluence of Yamuna channel, the braid bar deposits exhibit the growth pattern (Fig. 4.14).

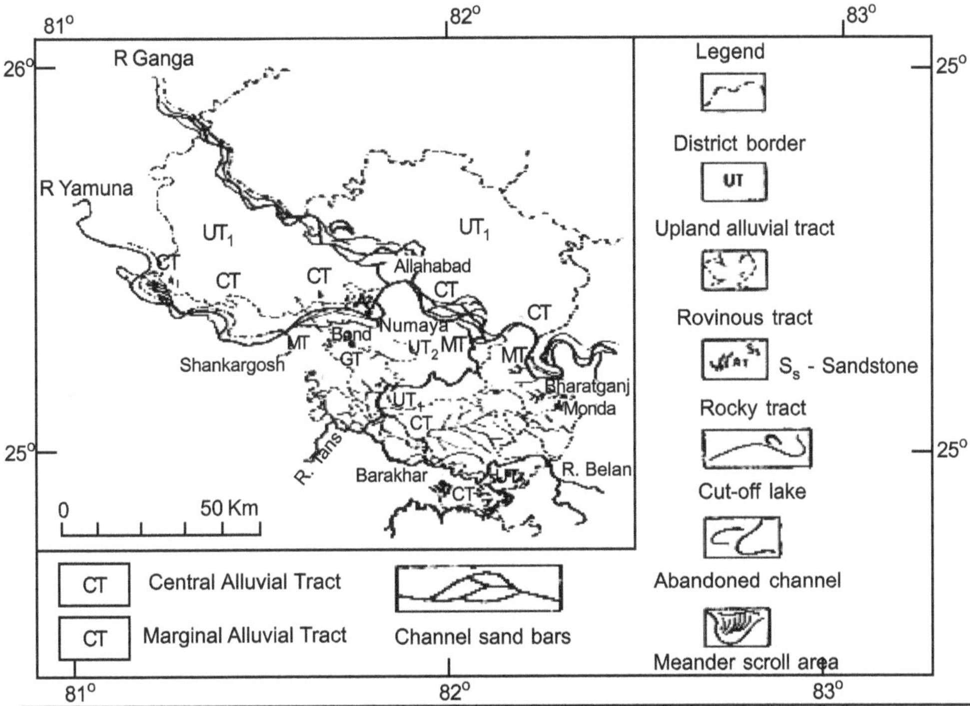

Fig. 4.14 Ganga and Yamuna channel patterns along with the geomorphic features. In Allahabad region as interpreted from satellite data

(Bajpai and Gokhale, 1986)

7. ***Alluvial fans :*** These are deposits occurring at the foothills of mountains grading into the plains. Mountain streams flowing in the valleys carry the eroded sediment, which is deposited at the exit point due to sudden lowering of the flow velocity caused by a sudden decrease of the gradient. The deposits are of broad conical shape with radial spreading of the sides all around (Fig. 4.15). They have concave upward profile with coarse material

Fig. 4.15 Alluvial fan.

(Courtesy : U.S. Geological Survey   Photo: Martin Miller)

on the upper slopes. Because they fan outwards from a point source, they are called alluvial fans. Their deposition, in addition to being initiated mainly due to the velocity, is also due to the infiltration of the water through the coarse material composing the fans. This infiltration arrests the transport of the fan material. Alluvial fans are common in arid mountain regions. Several adjacent fan deposits coalesce often to form a piedmont plain.

8. ***Delta :*** As the river approaches the sea, the flow velocity abruptly decreases as the slope of surface almost approaches that of the sea level. The sediment load rich in fine- grained sand, silt and clay is deposited. The coarser material is deposited at the mouth of the river forming the fore-set beds. The finer material settles down farther from the mouth as the bottom-set beds. The deposition of clay is facilitated by the flocculation of the particles in the brackish water environment. The water is brackish due to diffusion of saline water inland. The evolution of a delta is a balance between accumulations of fluvial material and its reworking, erosion and transportation by marine processes (Wright, 1985). As the delta expands, with more material getting deposited, the flow velocity of the stream falls rapidly and the already deposited sediment offers resistance to flow. The master stream in the deltaic region splits into several distributaries resulting in the extensive growth of a delta. Many of the widespread deltas with a large aerial extent have fan shape with the spread of the material outwards. Various subdivisions (Wright, 1985) of a deltaic region are indicated in Fig. 4.16. The subaqueous plain is composed of sands near the river mouths with clays towards offshore. The prodelta, composed of clays, is the last unit overlying a part of the continental shelf.

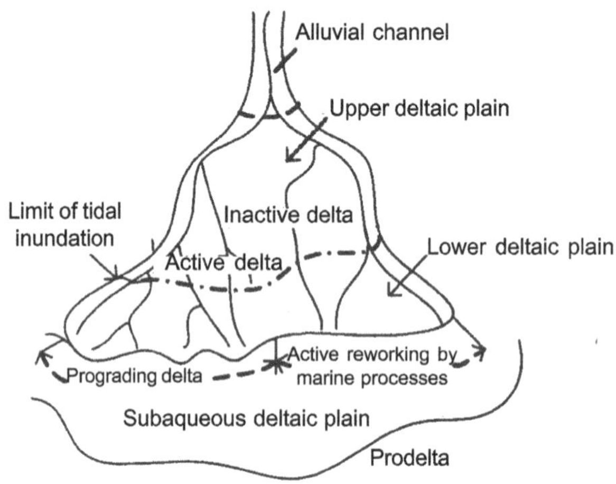

Fig. 4.16 Subdivisions of a typical deltaic region (Modified from Wright, 1985).

Deltas have different forms depending upon the spread of the river water (fresh water with sediment) into the saline environment with brackish water. If the densities of both are similar, then the fresh water with the sediment spreads out forming a fan shaped delta (Fig. 4.17). On the other hand, if the density of river water containing sediment is less than the seawater, the flow spreads in the form of a few channels, resulting in the formation of a delta resembling a bird's foot (Fig. 4.18). This type is also categorized as the river-dominated delta. If the density of the river water with sediment is more, it forms an elongated delta. The turbidity current with the sediment flows through the depressions or trenches to a long distance.

Fig. 4.17 Fan- shaped delta.

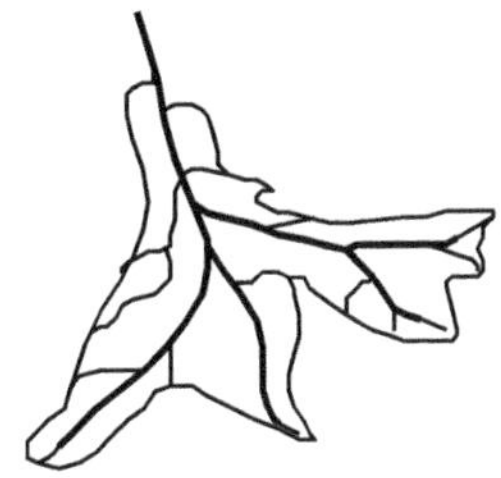

Fig. 4.18 Braid delta (bird's foot type).

The Ganga-Brahmaputra delta is the one of the largest in the world with an aerial extent of 105,640 sq km. The elevation varies from 15 m on the north to about 1m on the south with the surface sloping at .016 m/ km. It is bounded on the western side by the Lower Jurassic volcanic rocks and on the East by the sandstones and limestones of Eocene age. It is a result of combined delta-building activity of three drainages, namely Ganga, Brahmaputra and Jamuna networks. Around 4 billion tons of sediment load is annually contributed to Bay of Bengal by all these drainage channels. During monsoon period with the combined flow of about 82,000 Cusec, it behaves as a fluvial delta. The morphology reflects channel processes and the delta spreads outward into the basin. During the winter when the flow is about 10% of that in the monsoon period, the delta behaves as a tide-dominated delta. The sediments are brought back during high tides resulting in the redistribution of channel sediments forming sand-filled distributaries, carving low ridges and furrows. The eastern half of the delta (Fig. 4.19) is the active part of the delta. The inactive part is on the northeast. In this part, sediment contribution at present is almost absent.

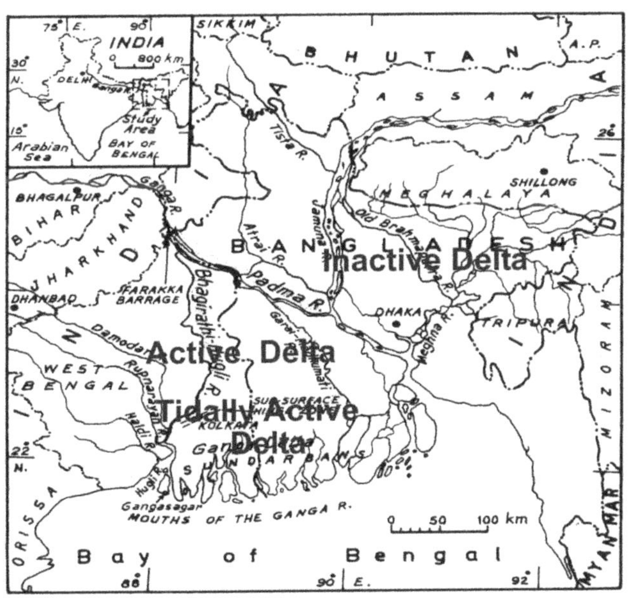

Fig. 4.19 The ganges delta.

Coastal Processes and Landforms

Coastline is the interface between land and sea. Erosion and transportation are facilitated by the wave energy. Waves are generated by the wind with the wind energy transferred to them. The train of waves created by wind vary with wind velocity, duration of the wind and also the area exposed to wind (also called as the fetch). The waves generated are due to a combination of all these three parameters. The waves vary in their form and wave height from small ripples common to lakes to huge tidal waves across the sea. Tidal waves generated during submarine earthquakes are known as tsunamis, which travel over large distances. Tsunamis are very destructive in nature.

Coasts are classified into three major types :

1. ***Primary coasts :*** These are generally rough and irregular in outline. Not much alteration by erosion is evidenced. The main processes operative include the modification of the coast by the changing sea levels and tectonic uplift as also delta development.
2. ***Secondary coasts :*** These are relatively older coasts modified by long-continued erosion and deposition of material. Typical erosional and depositional features commonly evidenced are the spits of different kinds, terraces, sea stacks (Fig. 4.20) and cliff arches.
3. ***Biological coasts :*** These are characterized by coral reefs and mangrove vegetation. The development of the coast is essentially through plant and animal growth. Coral reefs are common.

Fig. 4.20 Sea stacks formed at an erosional coast.
(Photo : Nagendra Kolluru)

In general, the coastal processes are due to the action of sea waves, generated off the coast, which approach the shore continuously. Wave motion involves movement of water particles at each place in orbits (circular to elliptic). The wave energy is transferred from particle to particle setting the wave in motion. Just like the elastic waves that propagate in earth materials, in water also the wave motion does not involve movement of the water body. It is only the waves that advance with the water particles staying where they are with their movement in orbits. The orbit diameter decreases with depth (Fig. 4.21), with the maximum at the surface. As the depth increases, the influence of wind decreases rapidly and hence the waves generated have a finite depth. As the waves approach the coast, when the depth of a wave is less than the sea bottom depth, the waves experience a drag of the bottom surface. As a result, the orbits of particle motion are elliptical instead of being circular (Fig. 4.21).

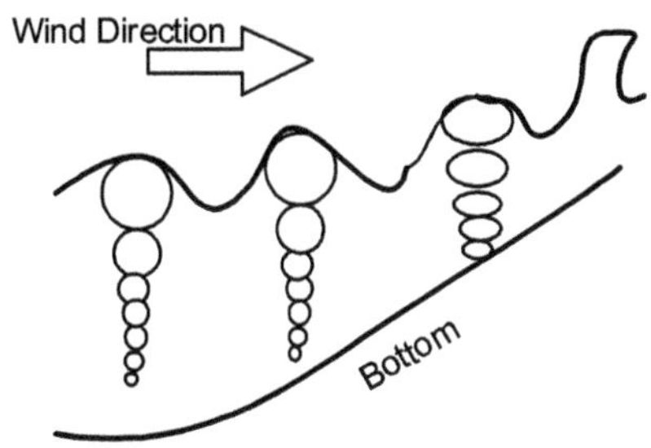

Fig. 4.21 Particle orbits in an advancing wave.

To define the geometry of a wave, the following terms are used : (Fig. 4.22)

- crest (highest point on a wave)
- trough (lowest point on a wave)
- wave height (the vertical distance between a crest and a trough)
- wave length (the horizontal distance between two adjoining crests)
- steepness (angle between a crest and the adjoining trough)
- period (time lag two crests passing at any location)
- celerity (C) which is a ratio of wave length (L) to the period (T)

  C is a function of depth (D) in shallow water while C is related to wave length (L) in deep waters.

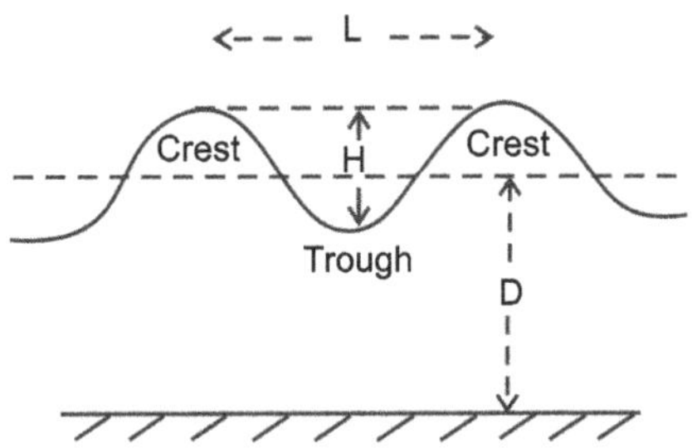

Fig. 4.22 Wave geometry.

Erosion and Deposition by Wave Action

Rolling waves, as they approach a shore, turn into breaking waves or breakers. At shallow depths (with depth of the bottom less than the wave length), the wave starts experiencing a drag due to the resistance to the particle motion by the interfering bottom surface. Since the period (T) remains same with the train of waves coming from the deep waters at the same rate, the wavelength gets shorter as the wave is slowed down in the shallow region close to the shore. The waves are consequently closely spaced with an increase in their height and steepness due to conservation of momentum of water. As the steepness increases, the wave base can no longer support the top and hence it breaks, consequently rolling down on to the shore. The zone in which the waves break is known as the surf zone due to the formation of surf (bubbly and foamy in character). The flow or rush of this surf up on the shore is termed as swash.

The return of this water into the sea is known as backwash. The backwash flows in the direction away from the shore and its angle to the shore is modified by the incoming waves in such a way that its resultant direction is almost parallel to the shore. The sediment load (sand) carried by the swash on to the shore is brought back by the backwash. This sediment carried parallel to the shore is known as longshore drift. Thus the longshore current is active in erosion and deposition of this sediment at different places along the beach. The shoreline is continuously modified by the longshore drift.

Wave refraction : Waves are rarely parallel to the coast. As they approach the shore, their incident direction gradually modifies because of the shallow depth. Part of the wave, which approaches the shallow depth, is slowed down earlier relative to the rest of it. This results in the bending of the wave approach direction. The incidence of the waves at an angle to the shore is known as wave refraction.

Wave refraction contributes to the erosion of headland projecting into the sea. Due to refraction, the waves approaching at an angle converge to the headland (Fig. 4.23). The erosion of the headland is due to concentration of wave energy. The material eroded is deposited in the adjacent bay (Fig. 4.24) where the refracted waves are divergent and have less energy. Thus in a coast with headlands and bays, the beach building process is active in the bay portions.

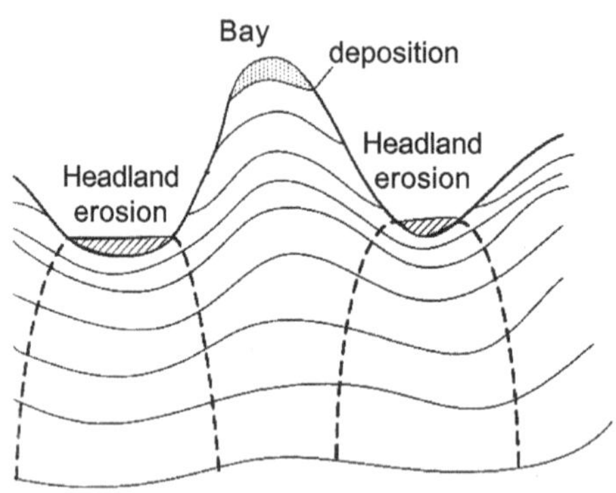

Fig. 4.23 Wave refraction and headland erosion.

Fig. 4.24 Headlands and bays along coastline in California.

Coastal Landforms

Wave-cut cliffs : Several of the erosional landforms are evidenced on coasts with headlands. The projecting rocky portion is leveled into steps or terraces known as wave-cut terraces or platforms. In addition, portions of the rocky headlands are separated from the main rock mass and stand out as isolated cliffs or sea stacks (Fig. 4.20) away from the shore These cliffs often exhibit arches, if the material is eroded through the cliff by wave action. These are typical features of an erosional coastline.

Beach : A beach is sand deposited part of the shoreline where the erosional and depositional processes are incessantly active by the waves and tides. A beach is subdivided into two zones – a foreshore and a backshore. The foreshore zone includes the sloping area occurring between low tides and high tides as well as the surf zone and the slope created by the swash. Its seaward limit is the breaker zone. The backshore covers the area towards the inland from where the swash zone ends up to the fore dune ridges. This zone is out of reach normally by either the ordinary waves or the tides. This zone has an influence of eolian processes towards the dunes. The extent of backshore modifies with seasons (Fig. 4.25). It is wider in summer season compared to that in winter. The region beyond the foreshore extending seaward up to the closure depths of waves is known as the shore rise. The nature of the beach material depends on the erosional inputs to the beach development. For example, in a rocky area, the beach has a large proportion of the very coarse material, often even coarse and rounded pebbles. Some beaches are flat and wide, composed of moist and fine sand material. The beach at Digha (in West Bengal close to Kharagpur) is an example of this kind and amenable to movement of vehicles including landing of small aircraft. In some areas, the sand dunes are built in the higher elevation of the backshore region.

A beach profile changes seasonally. Typical summer and winter profiles of a beach are indicated in Fig. 4.25.

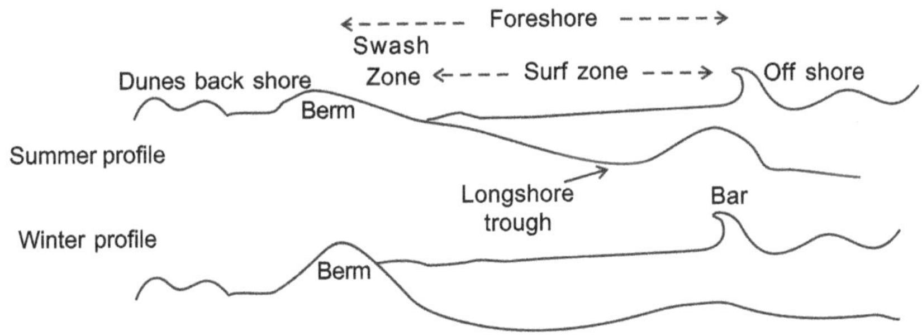

Fig. 4.25 Typical beach profiles.

Spits : These are linear deposits of sand extending out of a beach. If a spit blocks the entry to a bay, it is known as baymouth bar. Spits are composed of the sediment deposited by the longshore drift and as the sediment accumulates, it rises above the water level in the form of spits.

Barrier Island : A barrier island forms away from the main beach by the accumulation of the sediment drift away from the beach. A barrier island thus separates an open ocean from the near shore region. It is similar to a parallel beach and the water body in between is in the form of a lagoon with less of wave action.

Coral reefs : These are wave-resistant ridges or mounds built by corals and other marine organisms. Coral reefs develop in shallow warm saline waters (upto about 30 m of depth with temperature more than 20^{0}C throughout the year) where enough sunlight is available. If a reef has connection to the coastline, it is known as a fringing reef. A barrier reef is away from the coast with a lagoon in between. Coral reefs with circular outline around lagoons are known as atolls. These are formed on sinking coral islands. The coral reefs denote the sea level at the time of formation. If there is a subsidence of the coast, the reefs are submerged. On the contrary, their emergence out of water is an evidence for a coastline of emergence.

Estuaries : These are narrow bodies of seawater at the confluence of the river with the sea. Although an estuary is partially enclosed, it has a free connection to the sea (Fig. 4.26). The salinity of the lagoon water is controlled by the input of fresh water from the river and the diffusion of salt water into it from the sea. Depending upon the mode of formation of the barrier that partially encloses, estuaries are grouped as coastal plain estuaries, fjiord estuaries, bar-built estuaries and those formed by landslide mass flow. If there is more evaporation compared to water input, the salinity of water in an estuary is higher that of sea water, in which case it is known as a negative estuary. On the other hand, if the water is brackish (with salinity less than that of sea water), it is known as a positive estuary.

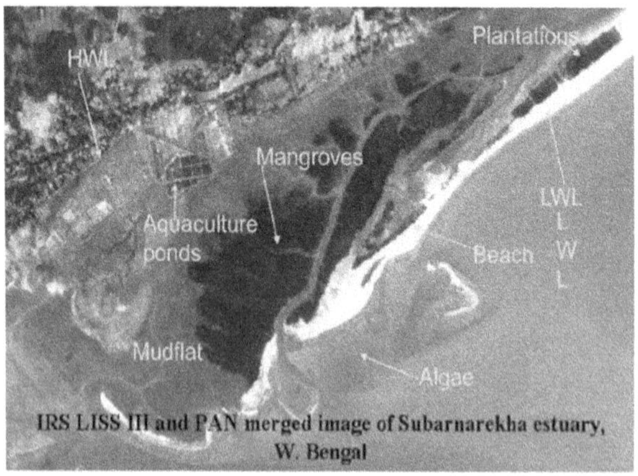

Fig. 4.26 Satellite image of Subarnarekha estuary (Nayak, 2003).

Delta : This depositional form has been discussed in detail under fluvial landforms in this chapter. Since tidal activity contributes to its development and morphology, delta is included also here in the list of coastal landforms.

Shore Protection Measures

Management of a coastline involves locating areas of erosion and implementing suitable measures for arresting the same. For the identification and demarcating locations where the coastline is being eroded, aerial photographs and satellite imageries are of immense use. Temporal data (imageries of the coastline taken at different periods of time) are to be utilized for this purpose. If the coastline photos or imageries (of the same scale) pertaining to different dates over a period of time are superposed, the extent and direction of erosion during that period can be delineated.

Once these areas are identified, shore protection measures can be planned and undertaken. Shoreline mapping has been carried for the entire Indian shoreline for the periods 1967-1968, 1985-1989 and 1990-1992 on 1:250,000 and 1:50,000 scale using Landsat and IRS data (Nayak et al, 1997). As an example, coastline changes in the area north of Chennai port are indicated in Fig. 4.27.

Coastal landform mapping has also been carried out for the entire coast on these scales (Nayak, 2002).

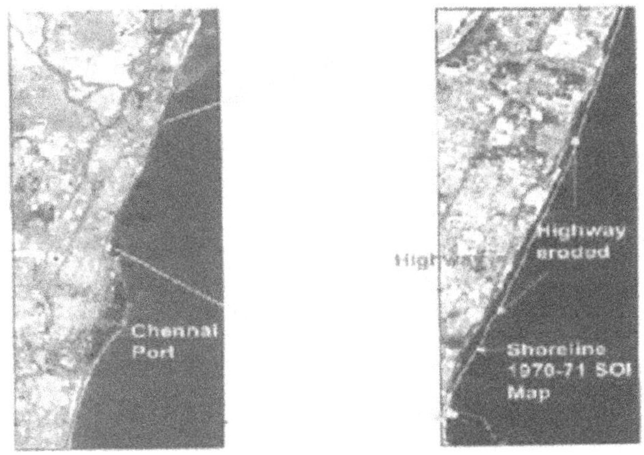

Fig. 4.27 Coastline changes in the area north of Chennai harbour Using IRS PAN data (Nayak, 2003).

Shore protection structures are of different types :

A. Engineering Structures :

Groins : These are engineering structures built perpendicular to the coastline (Fig.4.28) from the back beach into the sea up to the surf zone to check sand migration through longshore drift. The sand accumulated at the groins contributes to the development of the beach. Normally a series of groins (termed as groin field) are built to arrest the sand on the up-drift face side of the groins. However there is threat of erosion on the down-drift face of the groin, which can be minimized by suitably designing the groin spacings.

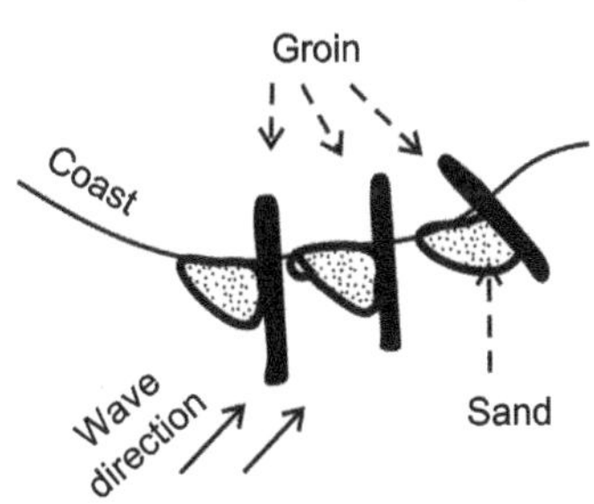

Fig. 4.28 Groins across a coastline.

Jetties : These are also perpendicular to the shore but built at the inlet points. To prevent the sand from blocking the inlet, these groins are built on either side of the inlet (Figs. 4.29, 4.30).

Fig. 4.29 California coastline with jetties.
(Courtesy: U.S. Geological Survey)

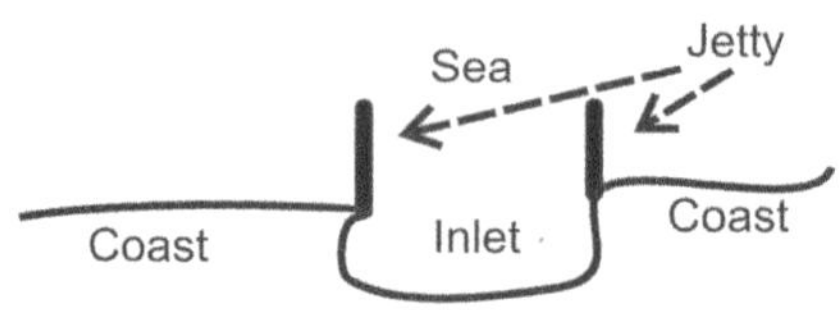

Fig. 4.30 Jetties built on either side of an inlet.

3. ***Offshore breakwaters :*** These are parallel structures to the shore (Fig. 4.31) built away from the shore (off-shore) to arrest erosion by absorbing the wave energy. Behind the breakwaters, sand accumulation takes place.

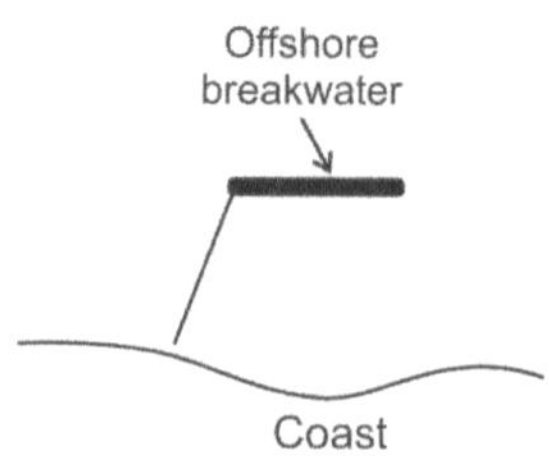

Fig. 4.31 Offshore breakwater.

4. ***Sea walls (Fig. 4.32) :*** These are structures built parallel to the coast to dissipate the wave energy and also to prevent inland flooding. They also help in preventing the slumping of cliff-eroded material.

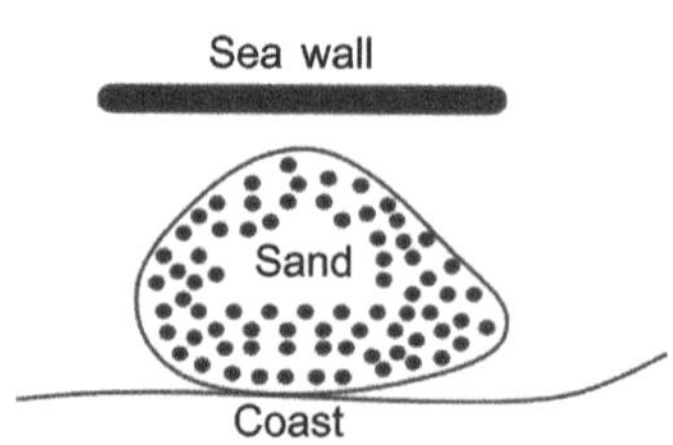

Fig. 4.32 Sea wall.

5. ***Revetments*** **:** Composed of stone or concrete facings, these structures are placed on the banks so as to absorb the wave energy (Fig. 4.33). These structures also protect the slope of a beach. They can be solid or porous (perforated) in nature. The perforated structure enables return of water filtered through it and retention of sand behind.

   A simple revetment consists of blocks of stones stacked randomly along the banks.

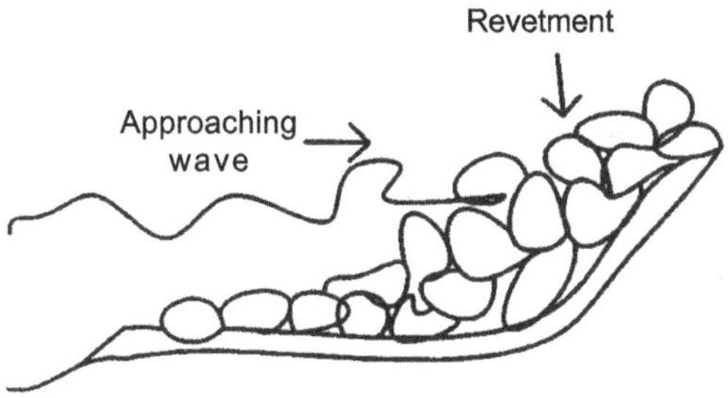

Fig. 4.33 Stone (boulder) revetment.

6. ***Bulk heads*** **:** These are vertical massive retaining walls for holding the sliding of material from the elevated portion of the shore into the sea. On low-energy coastline, gabion baskets (wire-mesh woven walls) are also used as bulkheads.

B. Non-engineering methods

1. Stabilizing the dunes and sand by planting vegetation
2. Beach nourishment through replenishment of sand at places of erosion

Glacial Erosion and Landforms

Glacial erosion and transportation are somewhat different from the common fluvial processes. Glacial movement is in the form of a basal flow or as a plastic flow. In the former, the entire glacier slips along a base caused by the melting at its base due to the enormous overburden of ice of the glacier over its base. This melt provides lubricating effect for the slip. In a plastic flow, the material moves as a viscous flow caused by movement of individual ice crystals resulting in the movement of the entire mass. Plastic flow is common in cold climates. The entire glacier is frozen and the plastic flow takes place in the upper part.

The eroding ability of a glacier is much more compared to that of a stream. It scoops and picks enroute-assorted load of different sizes. This assortment is composed of a wide spectrum of sizes from smaller particles to big boulders and of different compositions. Glaciers transport the load in bulk. During the transport, the material is eroded by dragging along the bottom. This results in parallel grooves or striations on the bed rock (Fig 4.34). These striations reveal the direction of movement of a glacier. The pebbles have triangular facets. Within the load, since the individual fragments of the material are held in place during the transport, they are not rounded as in the case of a fluvial sediment. The entire assorted deposit is known as a glacial till.

Fig. 4.34 Glacial deposit with polished surface, striations and grooves.
(Photo : F.E.Matthes Sequoia Courtesy : U.S.Geological Survey)

Boulder beds in a geological sequence are indicative of periods of glaciation. For example, the Gondwana sedimentary sequence starts with a basal boulder bed. The silt and clay fractions, deposited from the load on the bottom of a lake, exhibit a varved structure with alternating layers of fine and coarse-grained material. These occur as varved beds (Fig. 4.35) in this sequence, typical of glacial origin. Similarly the Siwalik sequence has boulder bed horizons indicative of cycles of glaciation. The recent glaciation was in the Pleistocene age. These boulder beds occurring in the field in consolidated and cemented form exhibit all the characteristics of a glacial deposit. Large boulders are known as erratic.

Fig. 4.35 Varved beds of glacial origin.
(Photo : P.Carrara Courtesy : U.S. Geological Survey)

The valleys carved by glacial erosion are U-shaped, in contrast to the V-shaped fluvial valleys. This is because the entire valley is filled with glacial load, which moves enmasse and carves a U-shaped valley with a broad base.

A moraine is a mound of glacial till dumped by a glacier. If it has accumulated at the end point of the glacial transport, it is known as an end moraine or a terminal moraine. On the other hand, if a glacier retreats due to warming and deposits its load, a recessional moraine forms.

Outwash is the sediment deposited by glacial streams flowing beyond the terminus of a glacier.

Drumlins are elongated hills composed of till and partly bed rock, developed by the flow of a glacier over a mound of sediment. The elongation is in the direction of flow. Commonly, drumlins extend over large distances.

Kames are small mounds of deposits formed through melting of a glacier at the margin of a receding glacier.

Eskers are long sinuous ridges formed by streams flowing in the vicinity of a glacier.

Kettles are depressions formed by melting of large blocks of ice left by a glacier long after the glacier recedes.

Eolian Processes and Deposits

Processes :

The processes of erosion, transportation and deposition by wind are known as eolian processes. Wind acts as a principal agent of erosion in arid climates. Transportation and deposition by wind agency is similar to the fluvial processes in that these are also controlled by the velocity, type of load and wind channel depth. Since viscosity is very low, the flow is turbulent. There is a limitation to the grain size of the load. The material is already weathered and in convenient sizes for transport by wind. However, enormous amount of load (sand and silt) is transported in desert regions. The eolian processes involve deflation, siltation and abrasion. Deflation involves plucking of the finer particles and lifting them. Siltation in the eolian processes is similar to that in fluvial processes. It involves lifting and displacing the particles to new positions in the direction of wind. Abrasion is the polishing and rounding of fine particles during transport by rubbing against each other. Sand carried by wind acts as an abrasive medium and modified the pebbles and other larger sizes on the ground. A wind-modified pebble, known as a ventifact, has typical carved and flat surfaces. Because of abrasion, the particles carried in suspension have frosted surfaces.

Deposits

Sand dunes :

These are ridges of sand deposited by wind. In the initial stages, sand carried as drift is accumulated due to low velocity. As the wind continues to flow, this drift material acts as an obstruction resulting in piling up of the load as a mound forming a dune. The wind erodes the sand dune forming a gradual slope on the windward side. On the leeward side, the slope is steeper and is known as a slip face (Fig.4.36). As several layers build up on the leeward side, the cross bedding develops. The cross bedding in eolian deposits is different from the one in fluvial deposits. In this case, the cross beds are truncated towards top and bottom and do not have asymptotic bottoms. Dunes migrate over large distances.

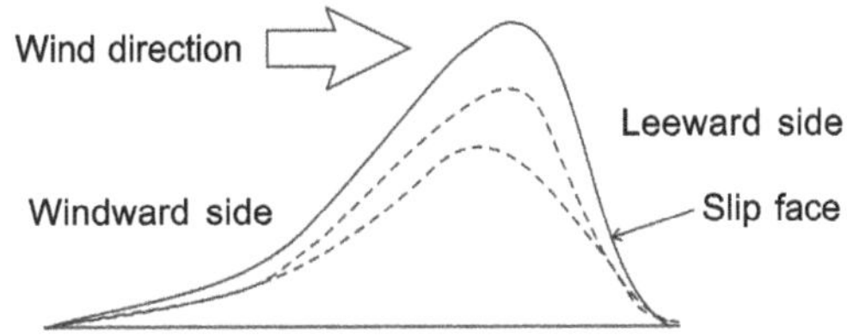

Fig. 4.36 Typical section of a sand dune.

Longitudinal dunes : These are long straighy symmetrical dunes formed parallel to the wind directions (Fig. 4.37). They commonly occur in cases where the material supply for transport is limited and strong winds blow from at least two directions. Slip faces occur on boh sides. These dunes extend over enormous distances.

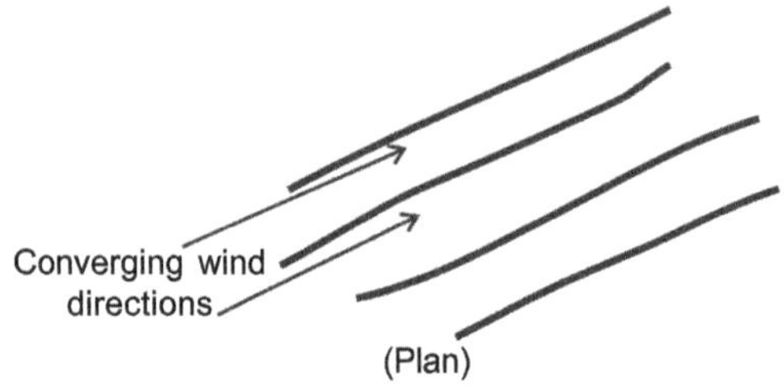

Fig. 4.37 Longitudinal dunes.

Transverse dunes : These dunes are asymmetrical and perpendicular to the wind direction (Fig 4.38). They form when enough material is available for accumulation as long ridges. The material in this case is evenly dispersed. They are typically formed on the back beaches by strong onshore winds.

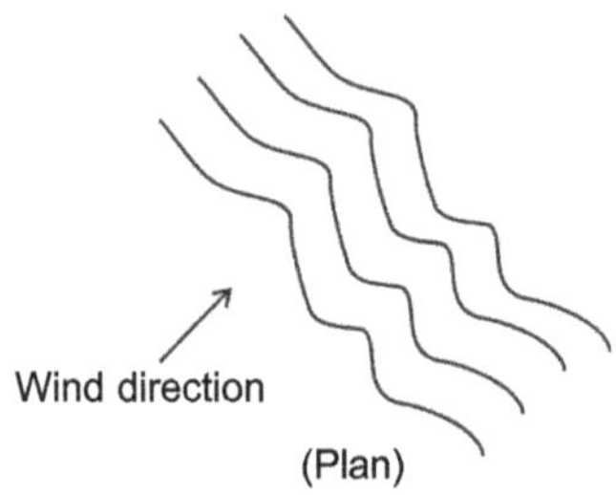

Fig. 4.38 (a)

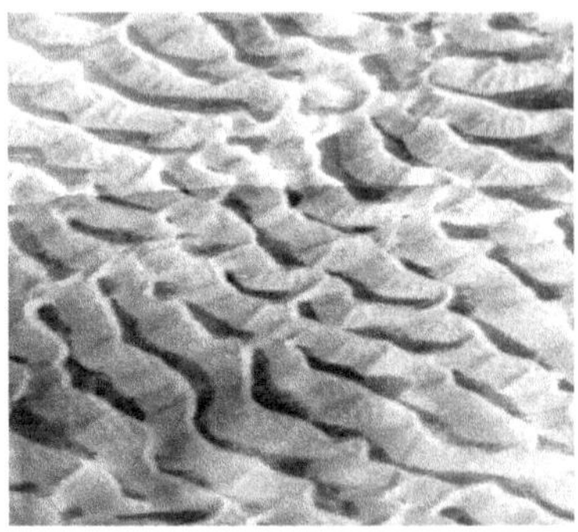

Fig. 4.38 (b) Transverse dunes
Photo: E.D. McKee (Courtesy : U.S. Geological Survey)

Parabolic dunes (Crescent dunes) : These dunes are common in semi-desert areas. When the material is blown out, the sides are blocked from movement by desert vegetation, resulting in an ultimate parabolic shape (Fig. 4.39). Multiple slip faces common with the slip face convex on the lee-ward side. The blowout forms are U-shaped.

Fig. 4.39 Crescentic dunes, Coastal Peru.
(Photo : John Mc Cauley Courtesy : U.S. Geological Survey)

Star dunes : These are dunes with radiating crests formed by winds from many directions. Typical star dunes are shown in Fig. 4.40. Maximum accumulation of sand takes place at the centre with arms radiating from it.

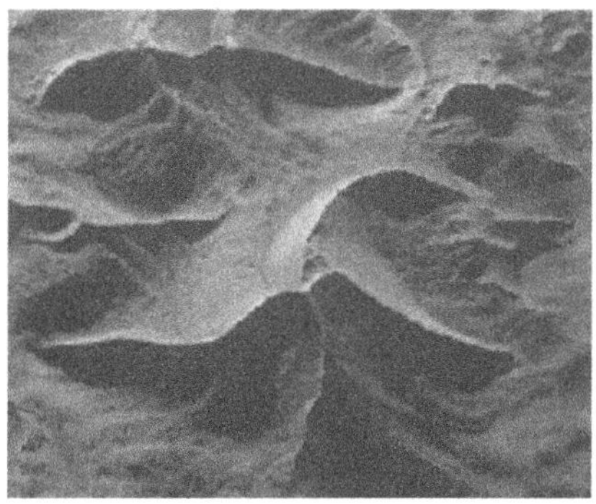

Fig. 4.40 Star dunes, Namib desert.
Photo : George Gester (Courtesy: U.S. Geological Survey)

Barchans : Very common in rocky areas. The material supply is limited for transport. During the migration of a dune of this type, the movement at the edges is much faster due to less material resulting in a crescent shape (Fig. 4.41). The slip face on the leeward side is concave. The points (wings) of the dune point downward and partially enclose the slip-face. The barchans occur in single units with single slip-face.

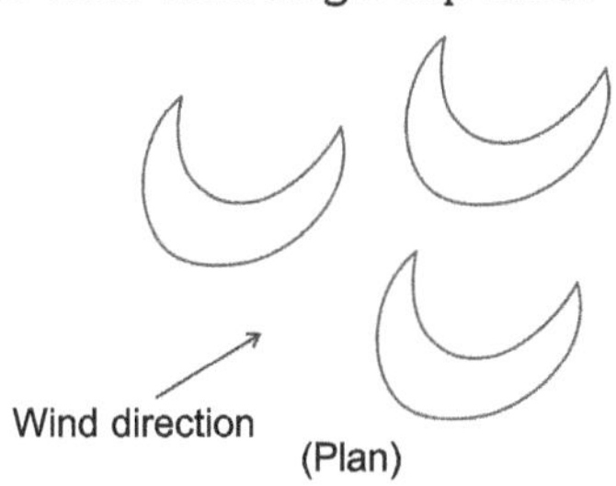

Fig. 4.41 Barchans.

Loess :

Loess is an eolian deposit composed of silt. It is characterized by its porous nature, and absence of layering. The silt particles exist in an interlocking texture. Loess takes steep stable slopes. In China, extensive loess deposits occur covering vast areas. The loess deposits are easily prone to erosion.

Engineering Problems in Eolian Deposits

Main problem with sand dunes is due to their migration. Before any construction is undertaken on dune sand region, stabilization of the dunes has to be accomplished first. This is accomplished by planting trees and special grass to hold the sand in place. For the construction of high-rise structures, pile – driving upto safe depths (to stable and compact underlying material) is undertaken.

Loess has the main problem of collapsing particle structure. It has a porous structure with fine clay matrix bonding the particles. Quite often carbonate material is also present. When loess is saturated with water, clay along with the silty particles of loess turns into a slurry, resulting in collapse (hydro-consolidation). This leads to failure of structures. In addition, piping (opening of paths for the water flow) occurs in loess. The channels formed by water continue to widen up leading ultimately to failure.

Desert Processes and Landforms

In desert environments, there is very little chemical weathering. Mechanical weathering is the dominant type. Deserts are characterized by scanty rainfall. Regular streams are rare. Flash floods occur. The streams, if present are of braided type. In humid regions, soil layer is present.

Desert Landforms

Pediments : These are broad surfaces with gently sloping rock formations, common along the foothill of the desert mountains. The highlands remain as isolated residual hills.

Inselbergs : These are steep-sided hills rising above the surrounding flat plains. These hills are composed of resistant rocks.

Playa lakes : These are intermittent or temporary lakes formed in depressions filled water in arid mountain regions. The water evaporates resulting in a dry lake. Since the water contains dissolved material, it is precipitated during evaporation and hence these dry lakes have salt concentrates (called evaporites).

5

Crustal Deformation and Geological Structures

The rocks in the Earth's interior are subjected to tectonic forces. These forces result in different stresses under different geological conditions on the rock material. These stresses may be uniform in all directions or oriented in particular directions. The pressure exerted by the confining material, being equal and uniform in different directions, is known as the confining pressure or the uniform stress. For example, the burial of sedimentary rocks at depth results in this type of stress. Similarly in metamorphism of regional type, the uniform stress plays an important role.

The directional stress can be compressional, tensional (often referred as extensional) or shearing type. The compressional stress results in the squeezing of the rock material while the tensional stress stretches or pulls apart the material. The shearing stress involves directional forces acting in opposite directions causing a slippage in the material. All these stresses in nature result in the development of geological structures of different types.

Stages of Deformation

When a rock is subjected to stresses, the resulting deformation is a function of the composition of the rock types, temperature, pressure and the strain rate. During the initial stages, the material behaves in an elastic manner during which period, the strain is reversible. Once the elastic limit is crossed, permanent deformation sets in, which on continued loading leads to failure or breakage of the rock material. Such rocks are known as brittle rocks. At low temperature, under low pressures and with high strain rates, the rocks in the earth's crust composed of hard minerals like quartz and feldspars behave in a brittle manner.

On the other hand, rocks subjected to high temperature, high confining pressure with low straining rates have shorter elastic span and the deformation is irreversible with continued loading. This type of behaviour is known as ductile deformation, characteristic of rocks at depth. Some types of rocks such as limestone (or marble) and sedimentary rocks rich in clay are ductile in their behaviour.

Ductile deformation is promoted by the changes at the grain level (intergranular) or at the lattice level (intragranular). At the intergranular level, due to the pressure at the contacts, partial dissolution of the grains takes place. Then readjustments in the rock material take place. On the other hand, the deformation at the lattice level is due to the movement (migration) of the lattice defects and vacancies, which increase with temperature and stress. Stylolites in limestones, flowage folds with contortions in shape, changes in grain boundary shapes and structures formed during the burial and compaction of sedimentary formations (diagenesis) are of this type.

The general stress-strain patterns typical of the brittle and ductile materials are shown in Fig. 5.1.

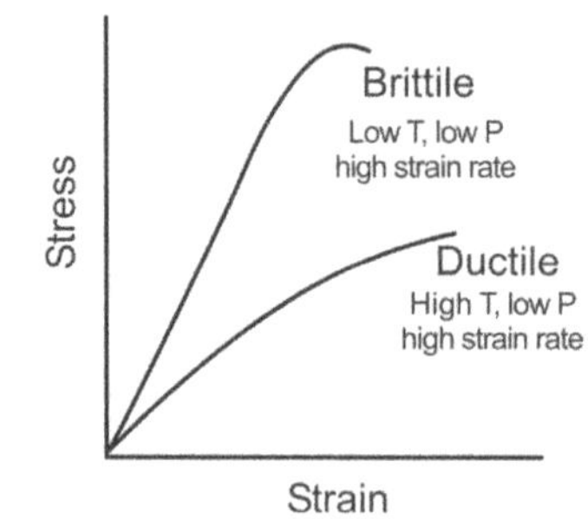

Fig. 5.1 Typical stress-strain patterns of brittle and ductile rocks.

The transition from brittle to ductile behaviour in earth's crust is known to occur at about 15 km of depth, beyond which the ductile behaviour is reported to continue up to the end of the crust. As the mantle region is reached, the rocks of the mantle are of peridotite type and being composed of high-density iron-magnesium silicates (olivine and serpentine) they are relatively of high strength. It has been established on the basis of available data that the rocks behave brittle in the upper part of the mantle to some depth beyond which the rock material softens leading to its ductile behaviour. This transition within the mantle is reported to be at around 40 km of depth from the surface (Fig. 5.2).

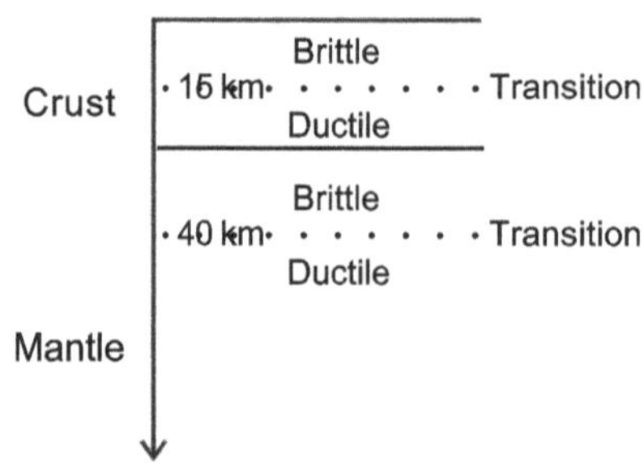

Fig. 5.2 Brittle and ductile transitions in crust and mantle.

Geological Structures

Geological structures are the evidences of crustal deformation. Depending on the process involved, particular types of structures develop in the geological formations.

A. Strike and Dip

Sedimentary formations at the time of their deposition are horizontal in most of the cases. Post-depositional processes tilt these formations and the angle of inclination to the horizontal is known as the dip. The direction of strike indicates the extension (or continuation) of formations. The horizontal line on a bedding plane indicates the strike direction. Dip is a vector quantity as it is represented both by the direction and angle of inclination. The strike and dip of formations are illustrated in Fig. 5.3.

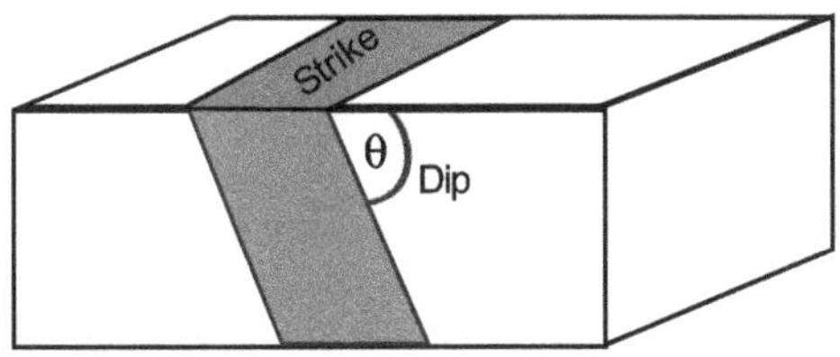

Fig. 5.3 Strike and dip of an inclined plane.

B. Folds

In the region where the rocks behave in a ductile manner, under stress they bend or fold instead of fracturing. At depth, the low strain rates promote the ductile behaviour.

Folds are structures that develop due to the compressive stresses acting for considerable periods of time. These are wavy patterns or flexures generated due to the plastic nature. Quite often these are associated with fractures indicative of their brittle characteristics. Tension joints or fractures develop on the convex side of these flexures.

Anticline and syncline (Fig. 5.4).

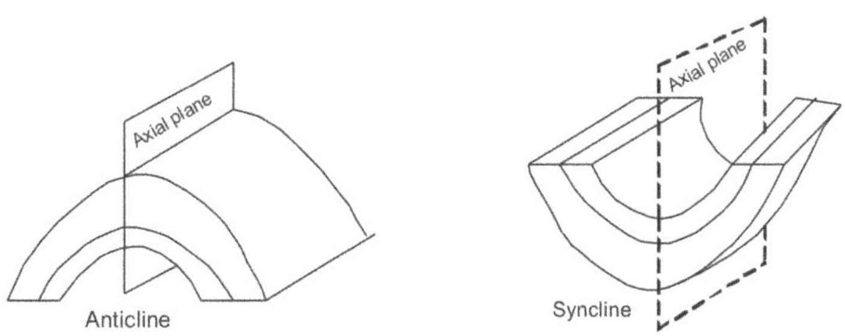

Fig. 5.4 Anticline and syncline.

If the folds are convex and are of arch shape, they are known as anticlines. If they have a trough-like appearance with concave shape, they are designated as synclines.

The sides of a fold are known as limbs. In the case of an anticline, the limbs on either side dip away from each other (Fig. 5.5). If a sequence of rocks is folded into an anticline, the oldest rock is at the core (or the centre).

Fig. 5.5 Anticlinal folding with tensional fractures and minor faults across the limbs at the crest. (Photo : Gokhale)

In the case of a syncline, the limbs on either side dip towards each other with the youngest rock located at the trough (or the centre).

A fold can be symmetrical or asymmetrical in shape. In a symmetrical fold, with respect to the axis (centre line) of the folds, parts on either side are symmetrical. The axis of the fold in such a case is vertical. The axial plane (a plane passing through the axis and the axial trace on the folded beds) is vertical (Fig. 5.4). The anticlines and synclines exhibiting these features are known as the symmetrical anticlines and symmetrical synclines respectively. On the other hand, in the case of asymmetrical anticlines and synclines, both the axis as also the axial plane are inclined (Fig. 5.6).

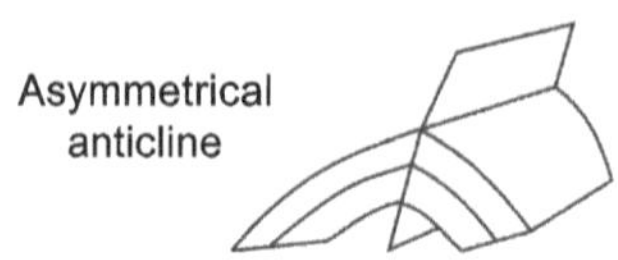

Fig. 5.6

If the axial plane is highly inclined with both the limbs dipping in the same direction, the fold is known as an overturned fold. An isoclinal fold is an overturned fold with limbs dipping in the same direction and with equal dip angles (Fig. 5.7).

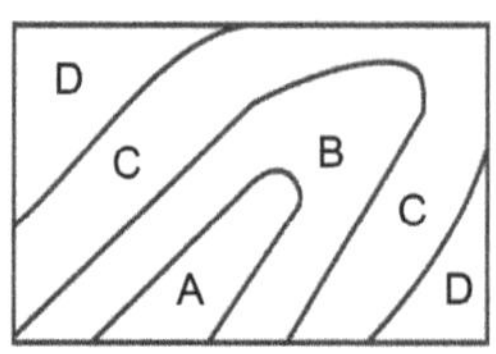

Fig. 5.7 Isoclinal fold.

Fig. 5.8 Tight folding in slates with slaty cleavage.
(Photo: A.M.Bhola, University of Delhi)

If the axial plane is horizontal, the fold is known as a recumbent fold (Fig. 5.9).

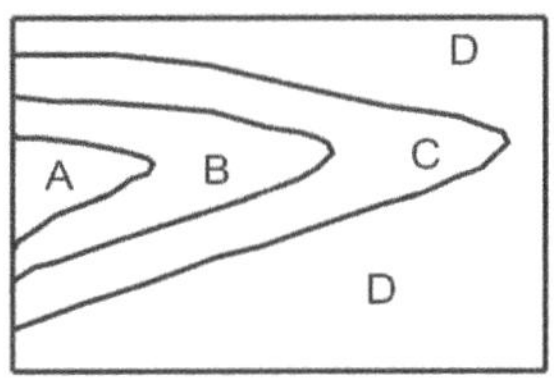

Fig. 5.9 Recumbent fold.

Folds exhibiting sharp v-like curvature are known as chevron folds (Fig. 5.10).

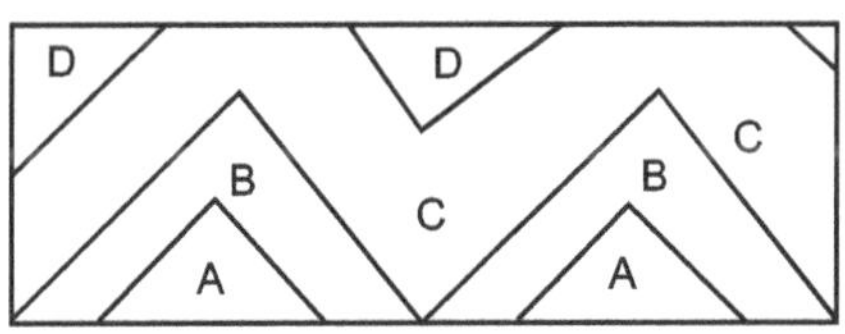

Fig. 5.10 Chevron fold.

Monocline is a fold with one limb nearly horizontal and the other limb sloping (Fig. 5.11). Monoclines have step-like appearances.

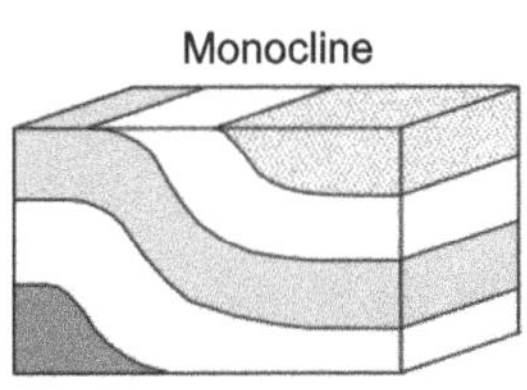

Fig. 5.11 Monoclinal folding.

Domes and Basins

A dome is a large circular or oval-shaped anticlinal structure sloping in all directions outward from the centre. A basin is a large circular or oval trough-like structure sloping inwards to the centre from all the directions. While lateral compression is responsible for anticlines and synclines, the domes and basins can also be formed by vertically acting stresses. A basin, for example, results from a constant subsidence (sinking) of an area over a long period of time.

Plunging Folds

These are the folds where the fold axis is not horizontal and is inclined. These folds close with distances and these closures are evidenced clearly in plan. In a plunging anticline, the fold axis dips towards the closure of the fold while in a plunging syncline, the fold axis dips away from the closure (Fig. 5.12).

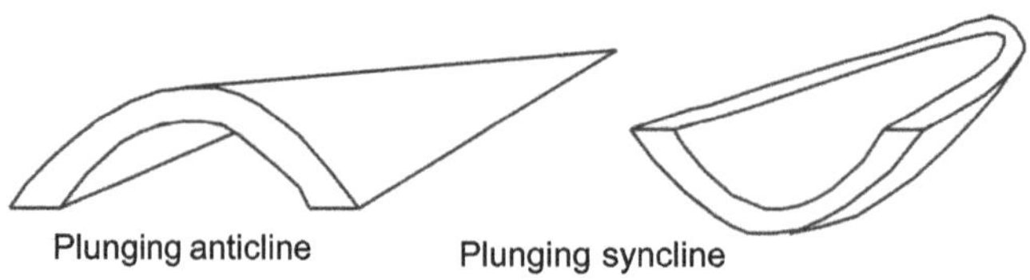

Fig. 5.12 Plunging folds.

C. Faults

Brittle formations subjected to stress deform resulting in the development of fractures. If displacement or offset takes place along these fracture planes, the structure is known as a fault. The plane along which displacement takes place is known as a fault plane. The side on which the formations have gone down is termed as the downthrow side. The vertical displacement of the same surface (which was continuous prior to faulting) on either side of the fault is known as the throw of the fault.

If the fault plane is inclined, it is a dipping fault. Just as in the case of the geological formations, the strike and the dip of a fault refer to the direction of extension of the fault on the surface and the angle of inclination of the fault plane to the horizontal respectively. The block, which is above the fault plane, is known as the hanging wall block while the one, which is below, is called the footwall block.

Faults, which are not active during a period of time, are known as dormant faults. Dormant faults get rejuvenated and fresh slippage occurs often during reloading of stresses and due to earthquakes at any later period of time.

Normal fault

A normal fault develops in brittle rocks subjected to tensional stresses. In the case of a normal fault, the hanging wall block moves down relative to the footwall block (Figs. 5.13, 5.14)

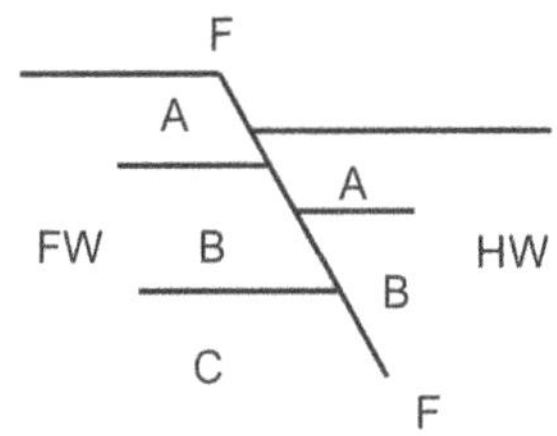

Fig. 5.13 Normal fault (HW: Hanging wall, FW: Foot wall).

Fig. 5.14 Normal fault across a quartz vein in foliated amphibolite.
(Photo : A.M.Bhola, University of Delhi)

If the central portion of an area within two normal faults goes down, the resulting structure is known as a graben (Fig. 5.15). On the other hand, if the normal faults slope away from each other and the central portion goes up, the structure is known as a horst. Grabens extending over large distances form the rift valleys. The famous rift valley of Africa is an example of this type. Horsts continuing over distances form linear mountain ridges.

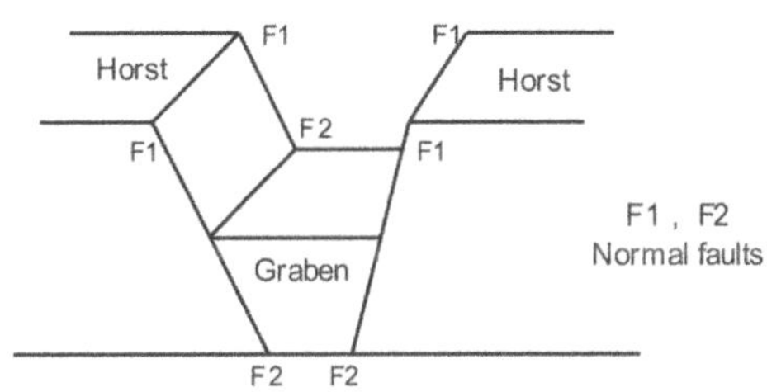

Fig. 5.15 Horst and graben.

Reverse fault (Fig. 5.16)

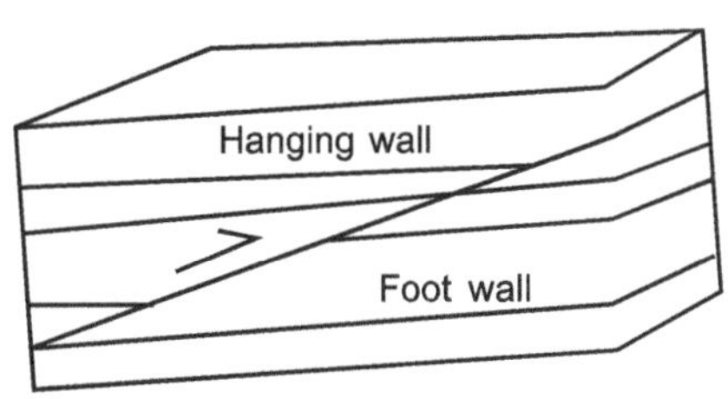

Fig. 5.16 Reverse fault.

A reverse fault develops in brittle rocks subjected to compression. In this case the hanging wall block moves up in relation to the footwall block.

A thrust fault is a reverse fault with the fault plane dipping at low angles, around 10^0 -15^0. Thrust faults extend over hundreds of kilometers. The older formations are thrust over the younger ones and move for considerable distances. In the Himalayan region, a series of thrust faults exist. The occurrence of older formations over the younger ones at higher altitudes in the Himalayan Ranges is the result of series of thrust faults.

Faults are also classified based on the movement as related to the strike or dip direction of the fault :

Strike-slip fault

This type of fault is formed due to action of shear stresses. The slippage takes place in a horizontal direction along the strike of the fault. Depending on the direction of displacement, the strike-slip fault is categorized as a left-lateral strike-slip fault or a right-lateral one (Fig. 5.17).

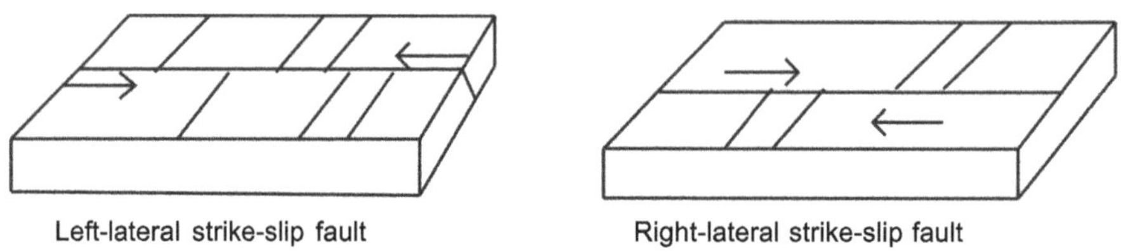

Fig. 5.17 Strike-slip faults.

In the case of a left-lateral strike-slip fault, to a person observing from one side of the fault plane, the block on the other side is towards the left. If it is towards the right, it is designated as a right-lateral strike-slip fault.

A transform fault is a type of strike-slip fault. Transform faults are common at the tectonic plate boundaries when they slide past each other. These are fracture zones confined to oceanic ridges, causing displacement of these ridges.

Dip-slip faults

If the displacement is along the direction of the dip of the fault, it is known as a dip slip fault (Fig. 5.18).

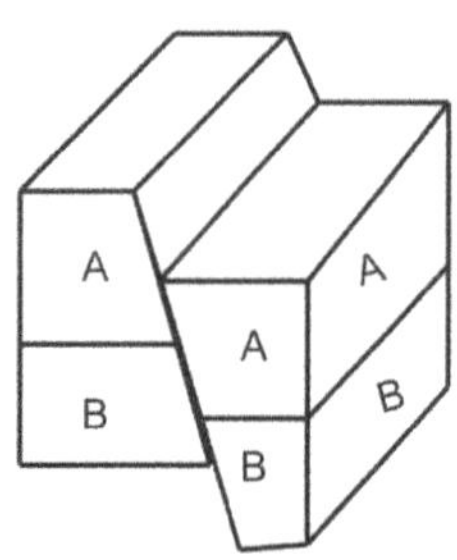

Fig. 5.18 Dip-slip fault.

Oblique-slip faults

If the slippage is at an angle to the strike and dip directions, it is known as an oblique-slip fault.

Stress orientations in faulting

Generation of faults of a particular type depends on the principal stress directions operative in the system. The maximum principal stress is σ_1. The intermediate and minimum principal stresses are designated as σ_2 and σ_3 respectively. If σ_1 is vertical, normal faults are developed. These fault planes dip at high angles. In the case of thrust faults, σ_2 is in the horizontal direction with σ_3 oriented vertically. If σ_2 is acting vertically, strike-slip faults are developed (Figs. 5.19).

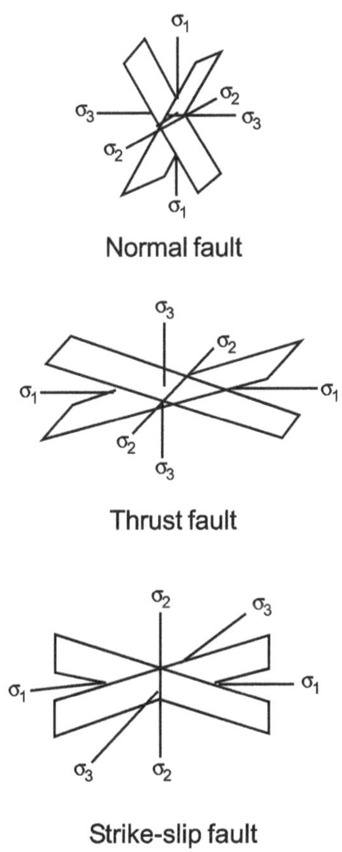

Fig. 5.19 Principal stress directions in faulting.

The fault planes, as evident from the figures, intersect along the intermediate principal stress (σ_2) direction. Any change in shape that occurs in the material depends on the difference between each of the principal stresses and the mean stress (the arithmetic average of these three principal stress values). These differences are designated as the principal deviatoric stresses in the respective directions.

Evidences of faults in the field

Existence of faulting can be confirmed from the following :

1. ***Fault breccia, fault gouge and mylonite :*** The rocks are crushed and the broken angular fragments are cemented in an altered matrix, which is termed as fault breccia. Due to intense frictional heat generated during the displacement, the formations are locally molten, forming a grey-coloured altered material, known as fault gouge. This is common in faulted shale or clayey rocks. If the material exhibits laminated structure with constituent grains deformed and partially oriented, it is known as mylonite. Mylonite is characteristic of deformation under intense temperature and pressure conditions.
2. ***Slicken-sides :*** These are step-like striations along the bedding planes of the formations (Fig. 5.20). They have a smooth feel in the direction of movement, while in the opposite direction they offer rough feeling, when felt by hand.

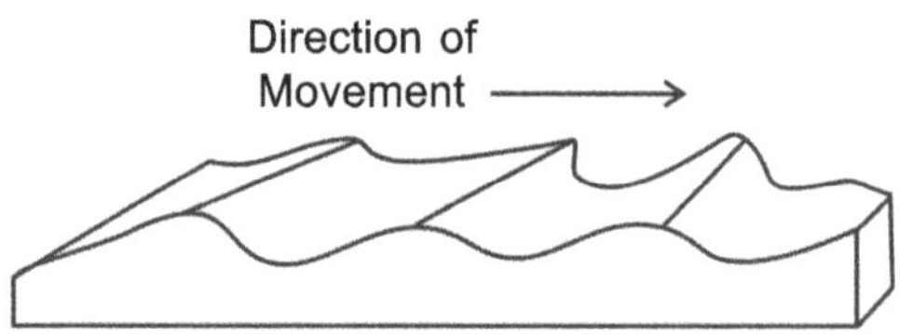

Fig. 5.20 Slicken-sides.

3. ***Springs :*** If the faults extend to depth, they provide channel ways for the escape of hot water resulting in thermal springs. If the faults intercept piezometric head even at shallow depths, groundwater springs result on the surface.
4. Dislocation of strata and occurrence of an older formation against a younger one with the strata in between missing.
5. Repetition and omission of strata.
6. Local disturbances of dips of formations
7. Dislocation of drainage
8. Sudden change in a drainage course, which follows the fault. In such cases, the drainage is said to be structurally controlled.
9. Streams cutting across competent narrow hill ranges. In the field, on both sides of the stream offsetting of the ridge is noticeable.
10. Occurrence of fault-scarps.
11. Occurrence of narrow valleys or zones of deformation.

D. Joints

Joints are fractures or openings in the rock formations. These differ from the faults in that there is no displacement along them. They are produced during the deformation of a brittle rock subjected to compressive, tensional or shear stresses. They can also be due to non-tectonic processes. The cooling cracks, due to tension, result in the columnar joints in basalts. The shrinkage during drying of argillaceous rocks result in mud cracks. Joints develop at the crests of anticlines due to tension. Shear joints appear in a deformed rock.

Just as in the case of geological formations, strike and dip of the joints are measured. Joints are classified on the basis of their orientation in relation to strike and dip direction of the host rock. If the strike of the joints coincides with the strike direction of the host rock, they are designated as strike joints. If they are at right angles to the strike of the host rock, they are known as dip joints. Joints in other than these two directions are grouped as oblique joints (Fig. 5.21). Joints in a formation can be present oriented in more than one direction. Joints with identical orientation are grouped into one joint set. Joint sets in an area are indicative of the direction of stresses operative in deformation. Thus they are useful in the kinematic analysis.

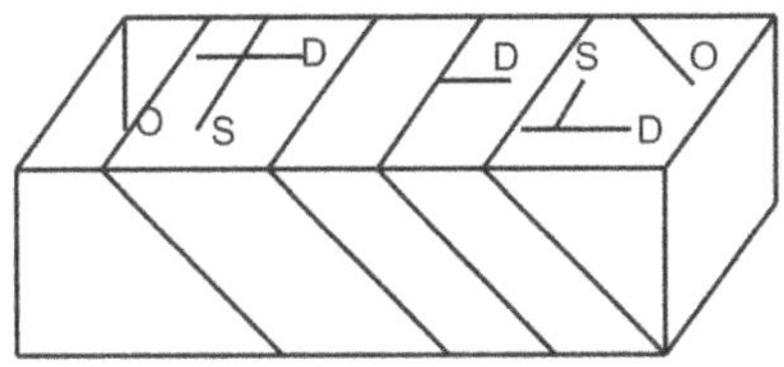

S : Strike joints D : Dip joints
O : Oblique joints

Fig. 5.21 Types of joints.

Quite often, narrow veins of quartz or other mineral veins of a later origin occupy these weak joint planes.

E. Unconformities

An unconformity represents a long interval of non-deposition during which erosion takes place. After a series of formations are deposited, they are uplifted and subjected to erosion. The pebbles or the eroded portions of the top formation are cemented together and occur as localized patches of conglomerate. Again the deposition of the next series takes place over the earlier sequence. The plane of separation between the earlier series and new series constitutes an unconformity. Thus the presence of conglomerate is an indication of an unconformity. For example, in the Vindhyan group of formations, conglomerate bed separates the two sequences of sandstone-limestone formations. The Cuddapah and Kurnool sedimentary sequences have conglomerates signifying the unconformities.

Unconformities are of different types :

Angular unconformity : The formations in the lower sequence have an inclination (dip) due to their tilting before the erosion event and then the latter sequence gets deposited. In such a case the unconformity is known as an angular unconformity (Fig. 5.22).

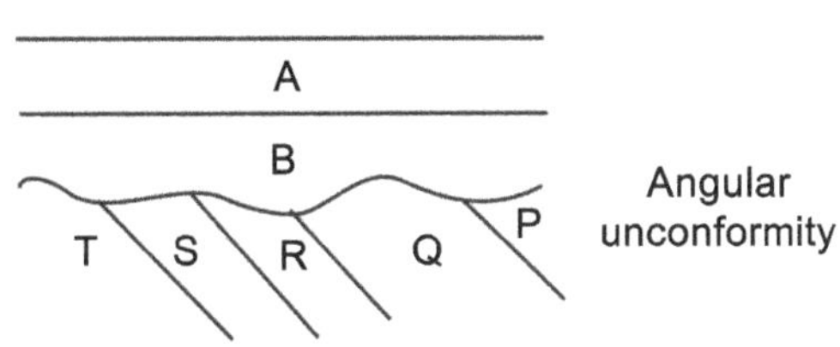

Fig. 5.22 Angular unconformity.

Disconformity : If formations in both the upper and lower series are parallel without any dips as in the earlier case, the unconformity that separates these two series is known as a disconformity (Fig. 5.23).

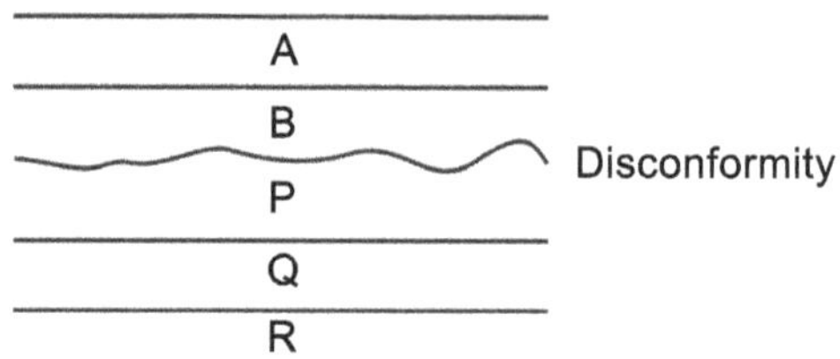

Fig. 5.23 Disconformity.

Nonconformity : If the sedimentary formations lie over a sequence of igneous or metamorphic rocks separated by an unconformity, the unconformity in such a case is termed as nonconformity (Fig. 5.24).

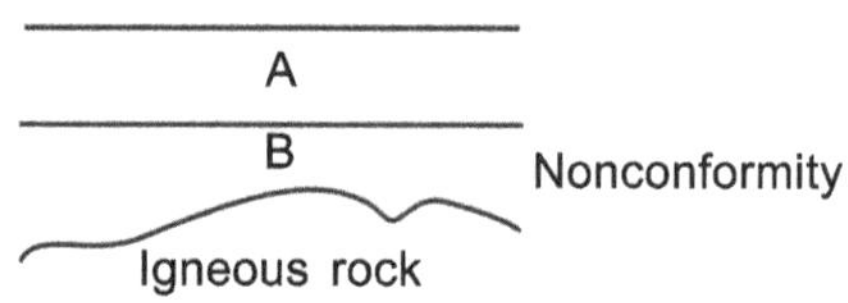

Fig. 5.24 Nonconformity.

Orogeny and Mountain-building

Mountain building is closely related to crustal deformation. There are several processes responsible for their origin. The normal and reverse faulting result in the uplift of areas in the form of fault scarps or escarpments. These stand out prominently as mountains, termed as fault-block mountains.

Compression due to the collision of continental plates results in the folding of the formations in between, ultimately leading to the development of mountain ridges. The uplift of Himalayas has been due to the collision of Peninsular Indian Block with the Tibetan Block.

Formation of mid-oceanic ridges at the divergent plate boundaries is due to pulling apart of these plates and formation of new crust with volcanic material issued through the fractures at the openings at the plate boundaries (Fig. 5. 25).

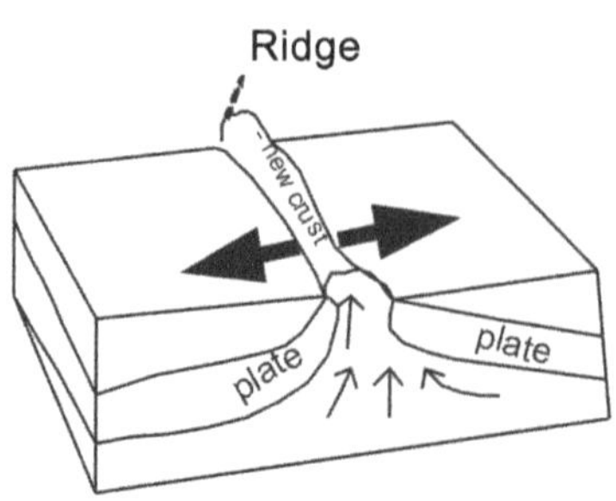

Fig. 5.25 Divergent plates and formation of ridge.

In the oceanic subduction zones, when plates collide, one of the plates goes under the other one (Fig. 5.26). This process is known as subduction . As the plate goes down, it melts and the resulting molten material comes out, through the cracks developed during the collision process, in the form of volcanic material, which accumulates on the surface.

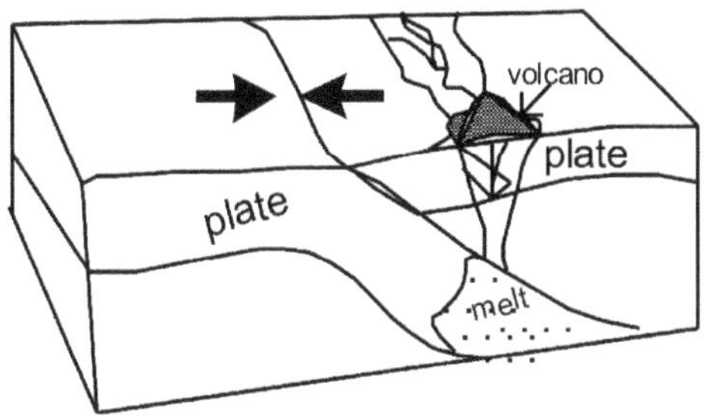

Fig. 5.26 Subduction zone (collision of plates).

If enough thickness accumulates, the solidified material emerges out of the ocean as island arc. The island arcs near Phillipines are a classic example of the same. All the island arcs have an arcuate (curved) disposition.

Mountains also can be due to reasons other than deformation. Volcanic eruptions on the surface and solidification of lava flows may lead to mountain building. The Deccan Trap hill ranges in Maharashtra are examples of this type.

Importance of Geological Structures in Civil Engineering Operations

In general, geological structures are an evidence of deformation in the rock formations. Whether they are joints, faults or folds, they are weak zones to be investigated and the appropriate treatment to be resorted to during the construction in a project. The formations are to be reinforced in terms of their strength. It is essential to prevent seepage in rock masses with discontinuities or planes of weakness through appropriate ground modification techniques in construction of dams, reservoirs and tunnels. The type of support and the extent of support to be provided for tunnels are designed on the basis of the structural characteristics and the strength of the rock masses. These are discussed in detail in chapter 9.

6

Site Investigations and Ground Improvement

Site investigations form an important initial step in any project. They are undertaken to obtain a final integrated surface and subsurface picture of the region under investigation. Site evaluation for a project, such as the location of a dam as well as the reservoir area, alignment of tunnels and evaluation of hazard zones as well as the construction of protective structures, involves investigations at the surface and subsurface levels.

Maps constitute an important segment of site evaluation. They range from regional maps to very detailed site maps. To cater to the specific requirements of the users, in addition to the topographic and geological information, specific relevant information is also incorporated on such maps prepared. For example, hydrogeological maps contain information not only regarding the aquifer potentialities of the formations but also information such as water table contours, water quality data, yields of wells at selected locations and indication of artesian regions

Engineering geological maps can be of different types. They can be with data meant for geotechnical and constructional requirements or for environmental management and protection. Maps with landslide hazard zonation or with depiction of endangered and risk areas due to natural disasters are very useful for disaster prevention and management.

Two types of multipurpose engineering geological maps are considered desirable (UNESCO-IAEG Guide , 1976) – one with engineering geological conditions (including rock types, landforms and variability as also the distribution of geodynamic phenomena) and the other with engineering geological zoning (based on the evaluation of spatial and functional relations among the basic geoenvironmental parameters).

Once the site is selected, competency of the geological formations in the site is examined. This includes not only the investigations in the site but also the laboratory tests of samples for strength and durability. Ground improvement techniques are adopted for reinforcing the strength, durability and for preventing seepage.

Site Investigation (Exploration) Methods

1. Geological
    - Field mapping
    - Remote sensing (aerial photos and satellite imageries)
2. Geophysical
    - Seismic refraction
    - Electrical resistivity
    - Geophysical logging
3. Subsurface drilling

Geological Investigations

These deal with the preparation of a detailed geological map based on (a) field mapping of formations along with structural details like joints, lineaments, shear zones, folds and faults as well as the weathered zones, (b) data from available aerial photos and (c) imageries prepared from the digital satellite data.

Topographic and geological maps constitute vital components of exploration. Topographic maps, as prepared by Survey of India, are available on different scales. While maps on 1:10,00,000 to 1:25,000 are utilized in the reconnaissance surveys of the region, detailed project site topographic maps are prepared through field surveying of the chosen sites. Topographic maps depict details including the elevation data (in terms of contours and also absolute triangulation values), drainage details (rivers and streams), surface water bodies (lakes and ponds), transport network (railway and road layout, bridges, culverts) and other conspicuous landmarks such as marked places of worship.

Identification of the relevant topographic maps for a region is done with the help of a guide map of the country prepared by Survey of India. On this map, the entire country is divided into different grids (starting from number 40 onwards). Each of these grids is further subdivided into 16 equal grids ("A" to "P"). The grids "A" to "P" are each further subdivided into equal grids numbering 1 to 16. Each topographic sheet at this stage is on a scale 1:50,000 and is indicated by a number (such as 56 F/12) on its top right hand side.

Topographic maps for any region can also be obtained on the basis of latitude and longitude values if the guide map is not readily available.

Field mapping : This involves transferring the geological and structural details observed on the site onto a base topographic map of a chosen scale. A geological map thus prepared depicts a spatial distribution of all the formations with their boundaries, lithological details (types of formations with compositional variations), and structural details in addition to the already existing features of the topographic sheet. Geological sections drawn across the map accordingly reflect not only the surface variations but also the continuity of formations, as extrapolated on the basis of surface details, to limited depth.

Remote sensing

Remote sensing involves the collection of data by indirect means (not in contact with the objects or phenomena under investigation) of detection, recognition and evaluation of the objects through distant sensing with the help of recording devices. Aerial photography and satellite remote sensing belong to this category. Data collected through these two techniques find extensive use in engineering geological investigations.

Aerial photos

During the early years of the introduction of aerial photography in exploration, black and white photos only were available. But with advent of time, colour photography, infrared photos and enhancement techniques for different features have increased the utility and popularity of this data. While black and white photos are suited for general mapping, colour infrared photographs are well suited for vegetation and water-vegetation discrimination. Aerial photographs are available on different scales. While for reconnaissance studies, small-scale photos (around of 1:40,000) are used, large-scale photographs of scales more than 1:10,000 are useful for detailed investigations.

Improved resolution and large-scale photography is possible through small format aerial photography. This is achieved through small format camera and Global Positioning System (GPS) fitted on a light aircraft. Aerial photographs are classified as vertical type or oblique type depending on the orientation of the camera. They can be either single lens or three lens photographs.

Aerial photo interpretation is based on the stereo viewing of an overlapping pair of photographs for any area. The photos have an overlap of 60% area and each has a side lap of 30%. The pair is viewed under a stereoscope for a 3-D view of the overlap by adjusting them in such a way under the stereoscope such that the areas fuse into one when viewed. Two types of stereoscopes are used namely the lens-type and the mirror-type. The former has two lenses fitted to a simple frame supported on short legs. The lenses can be adjusted to viewer's convenience for obtaining a 3-D view of the stereo-pair of the photos. In the mirror type, two mirrors are positioned at 45^0 to the horizontal with two prisms or two small mirrors also being mounted.

Following elements play a significant role in the photo-interpretation in terms of recognition of various features and patterns on the aerial photos :

1. ***Tone :*** It is a function of the amount of light reflected by the objects from the earth's surface and recorded by the camera. Light- colored material appears relatively lighter compared to a darker one. Materials (rocks and soils) with differing colour, composition and texture display different tonal contrasts on the aerial photos. Tone is also controlled by the type of the photo. For example vegetation which normally appears darker in tone appears relatively lighter on an infra-red picture. Nature and vigour of vegetation, soil moisture variations and subtle variations in the colour of the formations are very clearly observed on the colour images.
2. ***Texture :*** Texture is a result of tonal changes at different closely-spaced points of an object. Texture is very useful in the case of materials displaying similar tonal characteristics. It depends on the grain size, surface features and surface alterations. Texture is categorized as fine, medium or coarse type.
3. ***Size and shape of the features :*** Geological structures, geomorphic features and topographic features are identified from the characteristic shapes. Relative size variations indicate the magnitude of the landforms. For example the size of volcanic landforms, aeolian landforms and topographic highs and lows can be quantified in size.
4. ***Patterns :*** Drainage patterns can be easily classified into various types such radial, dendritic, rectangular and trellis. From the type of drainage, the permeability characteristics of the host formation can be inferred. For example, dendritic drainage is typical of relatively impervious formation. Landform patterns such as alluvial fans, beaches, tidal flats and bars appear distinct on the aerial photos.
5. ***Vegetation :*** Vegetation is often considered as an element in photo interpretation as the type and density of vegetation is indicative of the climatic conditions and the nature of the geological material. Variation of soil moisture conditions can easily be evidenced on aerial photos.

   An accurate model of a terrain can be developed by plotting all the ground points to scale with reference to three control ground points identified in terms of their geographical location on the photo and accurately coinciding the points on the photo with the base map. A stereoplotter is used for this purpose.

Remote Sensing through Satellites

Satellite data are extremely useful in terrain recognition and evaluation. Since a large region is covered, regional mapping is conveniently undertaken with the help of satellite data. From the data available in different bands, thematic maps can be prepared for features such as geomorphological details, geological structures, drainage patterns and topography. Additional advantage of remote sensing is that inaccessible regions can easily be covered.

Digital data obtained can be further enhanced for bringing out specific features using appropriate enhancement procedures. In engineering geological investigations, remote sensing data is useful in the site evaluation in dam construction, tunneling, groundwater development, landslide mapping and in location and alignment of transportation network. Data collected over a period of time (temporal data) for any area enable monitoring of changes during that period. For example, coastal erosion pattern can be monitored with temporal data resulting in the delineation of the active erosion locations on a coastline. This facilitates the appropriate choice and correct placement of the erosion protection structures along the coastline.

In satellite remote sensing, an object or a phenomenon is sensed by a sensor without its direct contact with the object. Since the object is remotely sensed, transfer of information to the sensor is accomplished through electromagnetic radiation. In a remote sensing system, the source of electromagnetic radiation transmits the radiation to the surface of the object. Although electromagnetic radiation covers a wide spectrum of energy of wave lengths ranging from 10^{-10} to 10^{10} µm, wavelengths of interest in remote sensing range from 0.3 to 15 µm. This range of optical wavelengths is subdivided into the reflective region, covering from 0.38 to 3.0 µm (spanning the visible from 0.38 to 0.72, near infrared from 0.72 to 1.30 and middle infrared from 1.30 to 3.00) as also the far infrared (thermal and emissive) between 7.00 and 15.0 µm. In addition to this optical wavelength portion, wavelengths between 1mm and 1m pertaining to the microwave region are used for microwave remote sensing.

These wavelength ranges are significant since different materials interact differently to these spectral ranges. Different materials (rock formations, soils, water, vegetation etc) reflect energy based on their composition, colour, moisture content, surface tonal characteristics and turbidity variations in water. In the visible and infrared ranges, the respective reflectances are recorded by the sensors. These sensors are placed on satellite platforms. In the case of thermal range of radiation, the reflected thermal energy is a function of the heat capacity of the materials and the thermal status (temperature). In the case of microwave spectral range, the electrical properties of the scanned material play an important role.

The first Earth Resources Technology Satellite (ERTS-1 or Landsat-1) was launched in 1972. Subsequently the others in the Landsat series (2 to 5) were put into operation. The periods of operation of all these American satellites are as follows:

Landsat - 1	July 23,1972 – Jan. 6 1978
Landsat - 2	Jan 22, 1975 – Feb. 25 1982
Landsat - 3	Mar 5, 1978 - Mar 31 1983
Landsat - 4	July16 1982 -
Landsat - 5	Mar. 1 1984 -

Three sensors, in combination were installed on these five satellites. These are RBV (Return Beam Vidicon) camera system, MSS (Multispectral Scanner) System and TM (Thematic Mapper). While the first two were on Landsat 1 to 3, Landsat-5 had MSS and TM sensors on board. The first two sensors (RBV and MSS) provided data on 4 bands and the data from TM was on 7 bands. Each of these bands corresponds to a particular spectral range of wavelengths and sensitive to specific features such as vegetation, water bodies, biomass in water bodies, vegetation, soil and rock types. The TM sensor has a spatial resolution of 30 m while the others in the Landsat system are with resolution of 79-82 m.

An earth observation satellite system was developed under the name SPOT by France along with Sweden and Belgium. The first satellite under this system was launched on February 21, 1986. The special feature of this facility is the stereoscopic scanning of the object from two different angles. The two sensors, MSS and PAN (panchromatic), installed have spatial resolution of 20 m and 10 m respectively and thus the data are very useful for detailed monitoring of resources and also for the planning of transportation and other utility services

A high-resolution satellite system known as IKONOS has been put into orbit by U.S.A. on Sept. 24, 1999. Among the existing satellites, this provides data with relatively high spatial resolution (of 1 m). Available in four bands within the spectral region of 0.45 to 0.90 μm, this data are well suited for operations such as urban planning and transportation network layout.

Indian remote sensing satellite systems

In India, the remote sensing satellite (IRS) activity has commenced with the launching of IRS-1A in March 1988. Subsequently IRS-1B, IRS-1C, IRS-1D and IRS-1E were put into operation. While the first three were launched by the Russian launching vehicles, rest of them (1D and 1E) have been launched with PSLV (Polar Satellite Launch Vehicle) which has been entirely built in India. The data from IRC-C with relatively good resolution (of 23.5m with LISS sensor) and being very cost-effective are very widely used all over the world.

In addition, IRS-P2, IRS-P3 and IRS-P4 (OCEANSAT-1) were put into orbit. These are extremely useful in producing data related to oceanography and vegetation dynamics.

The dates of commencement various Indian remote sensing satellites are as follows :

IRS-1A Mar. 17, 1988
IRS-1B Aug. 29, 1991
IRS-1C Dec. 28, 1995
IRS-1D Sept. 29. 1997
IRS-1E Sept. 20, 1993
IRS-P2 Oct. 15, 1994
IRS-P3 Mar. 21, 1996
IRS-P4 May 26, 1999

In addition, IRS-P5 (CARTOSAT-1) and IRS-P6 (RESOURCESAT) are also put into orbit. A Technology Experiment Satellite (TES) has been launched in 2001 to test advanced systems in remote sensing.

Details of the spectral ranges for different bands from IRS satellites are indicated in Table 6.1.

Table 6.1 Data from IRS satellites 1-A and 1-B
(With sensor : LISS : Linear imaging self scanning)

Band	Spectral range (μm)	Utility
1.	0.45 – 0.52	Coastal environment, forests, sedimentation
2.	0.52 - 0.59	Vegetation vigour, rock-soil contrast, turbidity
3.	0.62 - 0.68	Soil-rock boundaries, vegetation
4.	0.77 – 0.86	Land-water interface, geomorphic features

Data from IRS-1C also is from sensors PAN (panchromatic) and WIFS (wide field sensor).

The Indian National Satellite System (INSAT) was commissioned in 1983 for serving the needs in the areas of telecommunications, broadcasting, television, meteorology and disaster warning. This has also been put into excellent use for interactive education. Under this system, five satellites namely INSAT-2DT, INSAT-2E, INSAT-3B, INSAT-3C and METSAT (exclusively for meteorological purposes) are in operation.

Data interpretation

Interpretation is either visual or through digital data processing. Visual interpretation is carried out with imageries (pertaining to different bands). Different elements used in the recognition are :

1. Size : Dimension of the feature
2. Shape : Form of the object
3. Tone : Relative brightness of colour and contrast
4. Texture : Basic patterns characteristic of different objects
5. Site : Landuse , vegetated or forest land , water bodies
6. Association : Geomorphic features with different rocks
7. Drainage patterns

Digital image processing is carried out on computer. After the preprocessing corrections are made on digital data, the image is registered. Various enhancement techniques are used to bring out specific features such as lineaments, vegetated areas, and different formation characteristics. The image is classified in terms of various features by using either the supervised classification or unsupervised classification. In the former, training data sets specific to different features on the image are used for the classification. In the latter procedure, all the unknown data are classified into classes depending on the similarity in the data vectors.

The final classified image obtained either with visual interpretation or through digital data processing, is finally checked with ground truths.

With the introduction of Geographic Information Systems (GIS) as a tool, remote sensing data has gained wider applicability. GIS essentially is a database management system. It is a means of integrating spatial data of varying formats obtained on different scales and at different times Using the digital satellite data from remote sensing, individual data files are created for each of the involved parameters pertaining to any particular investigation (such as management of natural disasters like droughts, floods, mine and oil field fires, watershed management and transportation systems planning and analysis). Using GIS, an integrated picture is obtained through overlapping of the databases. Ranking the parameters in terms of their desirability, the entire region under investigation can be divided into zones with varying impacts. In addition to mapping and monitoring, GIS also helps in modeling, providing better alternatives to existing situations

In engineering geological investigations, remote sensing data are useful in the site evaluation in dam construction, tunneling, groundwater development, landslide hazard zonation, urban planning and in facility management including transportation routes. Data collected over a period of time (temporal data) for any area enables monitoring of changes during that period.

For example, coastal erosion pattern can be monitored with temporal data resulting in the delineation of the active erosion sites along a coastline. This enables an appropriate choice and correct placement of the erosion protection structures along the coastline.

Geophysical investigations

The geophysical methods deal with measuring of physical parameters of the earth's subsurface and interpreting the variations in terms of geological realities. In general, these techniques are divided into two groups:

1. Methods without depth control : Gravity, Magnetic
2. Methods with depth control : Seismic, Electrical

The first category are not useful for civil engineering practice since the measurement of either the acceleration due to gravity (g) or the magnetic susceptibility (μ) at different locations on the surface reflect the values due to the entire depth of the material present in the subsurface at each of these locations. As such there is no depth control. The magnetic and gravity methods are commonly adopted for assessment of material to depths of few kilometers as in the exploration for petroleum.

In Civil Engineering investigations, site investigations, either for foundations or for ground water, are limited to depths of the order of magnitude of few hundreds of meters (upto around 250 m). Hence the methods of the second category (seismic and electrical) find extensive use. Further, the depths of measurement can be controlled with appropriate layout of the equipment in the field.

Seismic methods

In seismic methods, two approaches are available: seismic refraction and seismic reflection measurements. Of these two, only seismic refraction method is convenient for Civil Engineering operations as it is simple in interpretation and best suited for shallow depths.

Seismic refraction method

This method deals with the propagation of seismic waves. The seismic waves generated due to a shock can be grouped under two types :

- body waves : P (primary or longitudinal) and S (transverse or secondary)
- surface waves : R (Rayleigh)

P and S waves propagate from the shock source in all directions in a spherical wave front. They penetrate the subsurface to greater depths depending on their energy. On the other hand, Rayleigh waves are surface waves with their propagation confined to near surface depths. The energy of a Rayleigh wave decreases very rapidly with depth. The R waves have large amplitudes and find extensive use in Soil Dynamics. Waves generated due to vibrations caused by the running of heavy machines or foundry forge plants etc are of this type. These waves, because of their large amplitudes close to the surface, are likely to cause damage to the structures in the vicinity of the vibration source. Isolation of the civil engineering structures from these vibrations is achieved by reducing the amplitude of these waves with the installation of trenches and sheet piles around the vibration source.

The P and S waves, although generated simultaneously due to a shock, travel with different velocities. The P waves travel faster than the S waves. In the site investigations for civil engineering purposes, the seismic refraction method involves measurement of the time of travel of either P or S wave from the shock source to the measuring location using a seismic timer (Fig. 6.1).

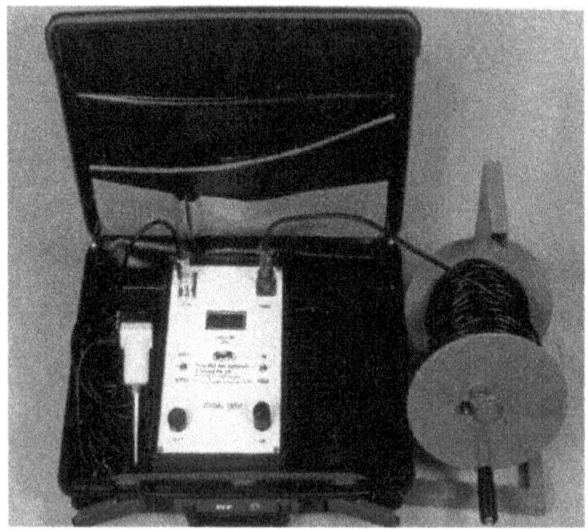

Fig. 6.1 Seismic timer unit and a typical geophone.
(Courtesy : Integrated Geo Instruments and services, Hyderabad)

Field methodology for seismic refraction method

1. Along a chosen traverse, geophones (also called transducers) are planted into the ground at regular spacings (say one meter apart) from a shock source (Fig. 6.2).

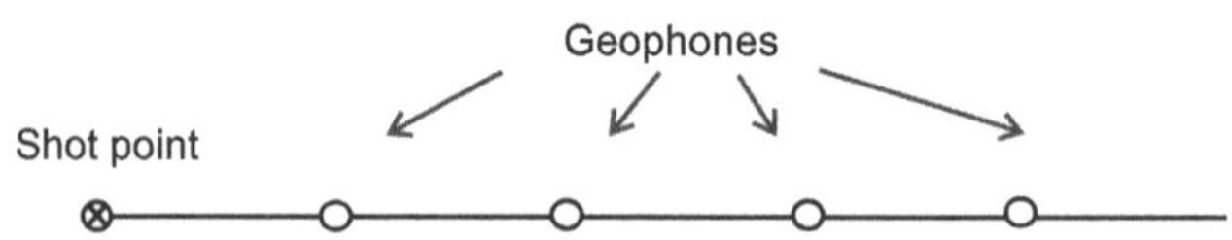

Fig. 6.2 Shot point and geophone lay out in a profile.

2. The shock is generated with a heavy hammer at the shot point.
3. The shock hammer and the geophones are connected to the refraction unit, which records the travel time of the wave as it arrives first at the geophone. This time is called the first shock instant.

   As the shock is generated, the P wave (or S wave) propagates in a wave front in all the directions. If we consider any path (trajectory) radial to the wave front, the time taken for the wave travelling along the trajectory close to the surface will be the first one to arrive at the geophones located nearer to the shock source. As the geophone distances increase, the wave travelling along the trajectory downwards and getting refracted at the interface will arrive faster at the farther geophones because of it's pursuing a part of its path in the high velocity material. (Fig. 6.3).
4. In the field, the shock is generated at the shock source and the time of arrival of the wave at geophone 1 (Fig. 6.3) is noted in the seismic unit.

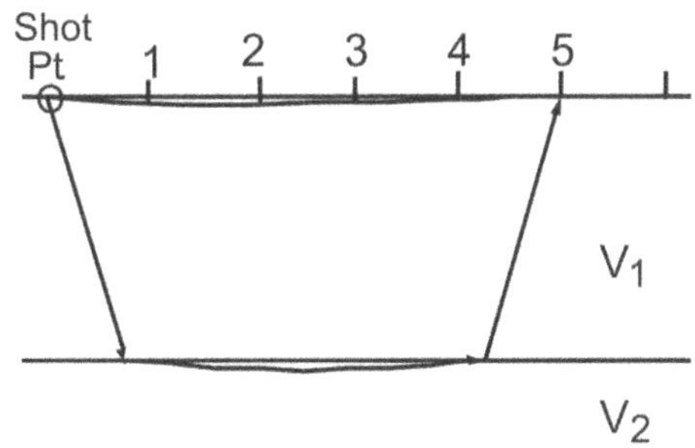

Fig. 6.3 Refraction of the seismic wave at the layer boundary.

5. The same procedure is repeated for measurement with geophone 2 and so on. If there is only one geophone available, measurements of time arrival of the wave are to be made with the geophone shifted to different locations (1, 2, 3 and so on as in the Fig. 6.3).
6. The measurements thus obtained from all these geophone locations are plotted against their distances of separation from the shock source .The resulting graph (Fig. 6.4) is known as the travel-time curve since it indicates the time of travel of the wave energy from the shock source to the respective geophones. It may be noted that the graph is linear with a break in its trend at some distance. The first segment starts right from the origin. The break in the curve is due to the wave arriving faster at the geophones located at distances beyond the break.
7. The measurements thus obtained from all these geophone locations are plotted against their distances of separation from the shock source. The resulting graph (Fig. 6.4) is known as the travel-time curve since it indicates the time of travel of the wave energy from the shock source to the respective geophones. It may be noted that the graph is linear with a break in its trend at some distance. The first segment starts right from the origin. The break in the curve is due to the wave arriving faster at the geophones located at distances beyond the break.

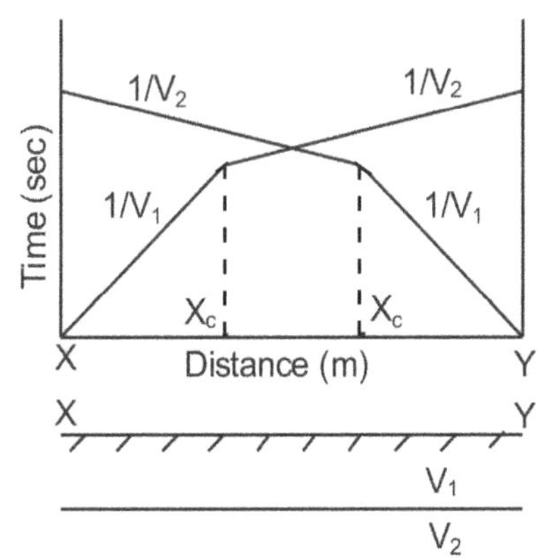

Fig. 6.4 Travel-time curves for two layers with refraction survey along a profile in forward and reverse directions.

Inverse of the slope of each of these straight-line segments corresponds to the respective velocity with which the wave travels in the formation.

The seismic wave velocity (both for P and S waves) varies with the material (Table 6.2). It is a function of the elastic constants and the density of the material. In soils and highly weathered formations, the P wave velocity is commonly in the range 500 to2500 m/s.

In hard rocks, the order of magnitude of velocity is of 5000 m/s. The velocity reduces if the rocks are jointed. With the formation saturated (below the groundwater table), the velocity improves in a formation. Thus using seismic refraction techniques, the nature and extent of the aquifer as also the other semi-pervious and impervious formations (aquiclude, aquitard and aquifuge) can be conveniently estimated.

Table 6.2 Ranges of seismic wave velocities

Formation	P-wave velocity (m/s)	S-wave velocity (m/s)
Massive granite	5000 – 7000	2500 – 4000
Weathered granite	800 – 2500	450 – 1500
Sandstone	1500 – 4000	1000 – 2500
Limestone	2500 – 6000	1600 – 4000
Sand	300 – 800	200 – 450

Utility of travel-time curves

1. ***Estimation of velocity values and identification of the formations :*** The velocity values are estimated from the segments on the travel-time curves .These velocity values are interpreted in terms of the geological formations in the subsurface. For this, a broad idea of the possible formations is necessary. Taking one profile at a place in the region where subsurface geological details are available will facilitate better interpretation.
2. ***Estimation of thickness of different formations :*** From the travel-time curve (Fig. 6.4), the velocities and the critical distance " X_c" of the break on the graph, the thickness of the layer is estimated as :

$$D = Xc/2. \sqrt{\{(V_2 - V_1) / (V_2 + V_1)\}}$$

3. ***Identification of dipping formations and estimation of dip angle :*** For formations without any dip (horizontal formations), the travel time curves obtained from profiles in the forward direction (from X to Y with shot point at X in Fig. 6.5) and in the reverse direction (in the direction Y to X with shot point at Y) will be identical in terms of the slopes of the first segments, slopes of second segments and the X_c values of both the travel-time curves (Fig. 6.5).

If the travel-time curves for the forward (X to Y) and reverse (Y to X) profiles are not identical (Fig. 6.5), the formations are not horizontal but are dipping. Only the slope of the first segments is equal but not that of second segments as also the "X_c" values of both the travel-time curves.

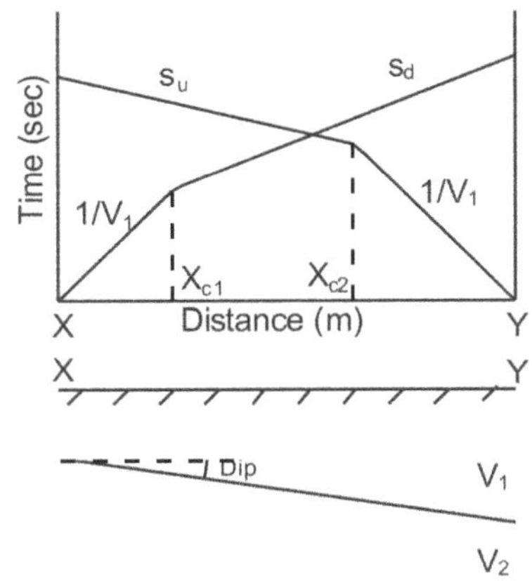

Fig. 6.5 Travel-time curve for formations with dip.

The dip angle (ϕ) is estimated from the velocity (V_1) and the slopes of the second segments (s_d and s_u in Fig. 6.5) as

$$\phi = ½ (\sin^{-1} V_1.s_d - \sin^{-1} V_1.\ s_u)$$

where s_d and s_u are the slopes of the down-dip and up-dip segments (Fig. 6.5) respectively.

4. ***Estimation of dynamic elastic parameters of the formations :*** This is done based on the primary wave and transverse wave velocities (V_p and V_s) of each of the individual formations as :

$$Vs = (\mu / \rho)^{1/2} = [(E / \rho).\ ½ (1 + \rho)]^{1/2}$$

$$Vp / Vs = (k / \mu + 4/3)^{1/2} = [(1 - \sigma) / (½ - \sigma)]^{1/2}$$

Where E = Young's modulus

μ = Rigidity modulus

k = Bulk modulus

σ = Poisson's ratio

ρ = Density

5. ***Detection of fault and estimation of the downthrow :*** In the case of a profile across a fault, the travel-time curve will have an extra segment between the regular first and second segments (Fig. 6.6) due to the delay time in the travel of the waves because of the formation displacement.

Downthrow of the fault (Dt) = $(t_2 - t_1).\ V_1 V_2\ /\ (V_2^2 - V_1^2)^{1/2}$

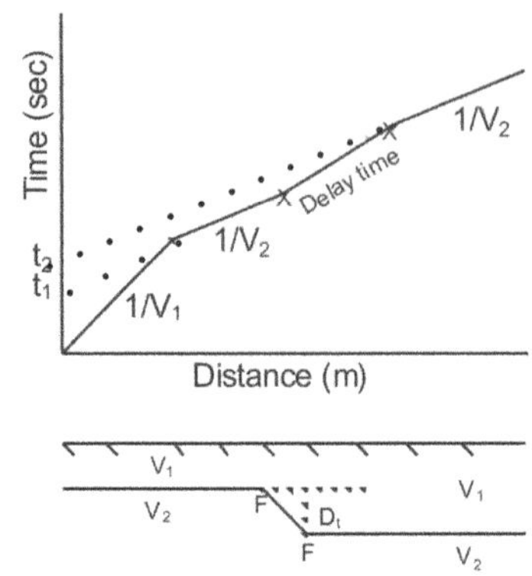

Fig. 6.6 Travel-time curve for formations with a fault.

The downthrow estimates from this procedure have been reported to be very accurate. During the site investigation in the Kopili hydroelectric project site in Assam (Datta *et al*, 1972), a fault was detected in the Eocene formations (Kopili shale and Sylhet limestone). This was evidenced from the extra segment in the travel-time curve. To check the validity of the downthrow estimates for three profiles at different places across the fault, drill hole data were obtained through drilling at critical places. It has been observed that the estimated downthrow values are in close agreement with the observed values (Table 6.3).

Table 6.3 Downthrow values in Kopili Project area

Estimated (m)	Actual (m)
53.88	54.0
34.64	37.0
65.52	66.0

(Source: Datta et al, 1972)

Electrical methods

Under this group, the following are commonly adopted in ground water investigations:

- Surface electrical prospecting
  Electrical resistivity method
  Self-potential method (S.P)
- Well- logging technique

Electrical resistivity method

Electrical resistivity is the inverse of conductivity. It varies in geological materials with composition, pore water saturation and nature of pore water (salinity). In clays and shales, since the pore water contains electrolytes, the resistivity is relatively low compared to sandstones or other formations. A lowering of resistivity of a formation with water saturation below the water table is clearly reflected in the field data. Further, a very low resistivity value is recorded in aquifers with saline water. This method is very popular for groundwater exploration. Typical resistivity ranges of common geological materials are presented in Table 6.4.

Table 6.4 Typical ranges of electrical resistivity values

Formation	Resistivity (ohm.m)
Sand with saline water	< 5
Sand with fresh water	80-200
Clay	1-50
Sandstone	100 - 200
Limestone	100-1000
Shale	50-200
Massive igneous rock	>3000
Jointed igneous rock	500-1500

This method involves inducing a potential in the ground and measuring the same. Four electrodes are used- two for passing the current and two for measuring the potential difference. The layout is schematically indicated in the Fig. 6.7.

The current electrodes (C_1 and C_2) are metal rods (of cast iron). The potential electrodes (P_1 and P_2) are non-polarizing type. Cylindrical porous non-glazed ceramic pots with a rubber stopper and a thin copper tube are used for this purpose. The ceramic pot is filled with super saturated copper sulphate solution with the copper tube dipping in the same. Non-polarizing electrodes are used for potential measurement to counteract the mobility of ions (copper) from the positive to the negative side (P_1 to P_2) when the ground is energized and potential is induced. If the non-polarizing electrode set up is not used and instead the copper rods are directly planted into the ground, no steady reading can be obtained because of the polarization effect. Instead of such non-polarizing electrodes, stainless steel rods are also used as potential electrodes.

The field procedure involved is as follows :

1. The four electrodes are planted into the ground with the spacing between them chosen depending upon the field situation. Two types of electrode spacing are commonly used in ground water investigations (Fig. 6.7).

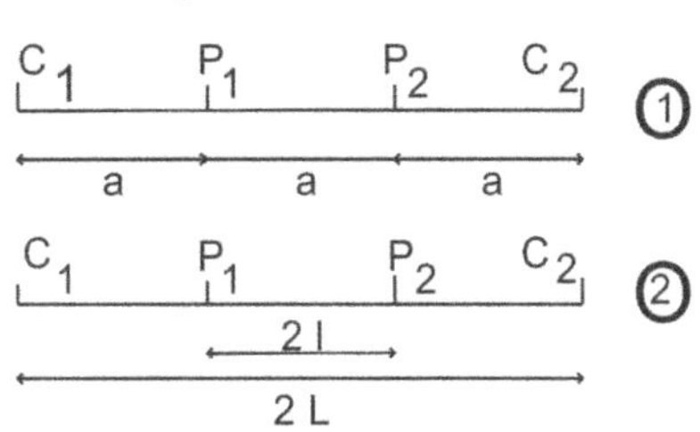

Fig. 6.7 Electrode spacings -1.Wenner, 2. Schlumberger.

In the Wenner's spacing, the distance between any two adjacent electrodes is kept same. This spacing is commonly designated as "a". However in the Schlumberger's spacing, the distance between C_1 and C_2 is designated as 2L and the potential electrodes P_1 and P_2 (with separation of 2L) are placed symmetric to the centre of the profile (Fig. 6.7). Commonly, Wenner's spacing is adopted in the field for its simplicity. With this spacing, the computed resistivity values for different spacings correspond approximately to depths equivalent to the spacings.

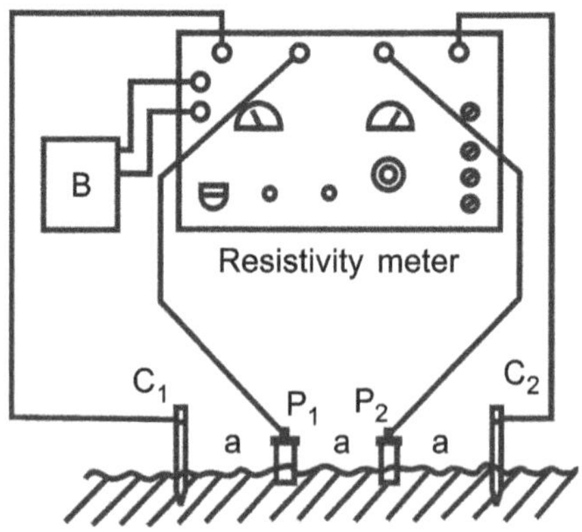

Fig. 6.8 Schematic circuit diagram for electrical resistivity technique (C_1, C_2 : current electrodes : P_1, P_2 potential electrodes; B : battery).

2. The two current electrodes are connected to C_1 and C_2 of the resistivity equipment (Fig. 6.9). Similarly the potential electrodes are connected to P_1 and P_2. The D.C. battery pack is connected to the respective terminals.

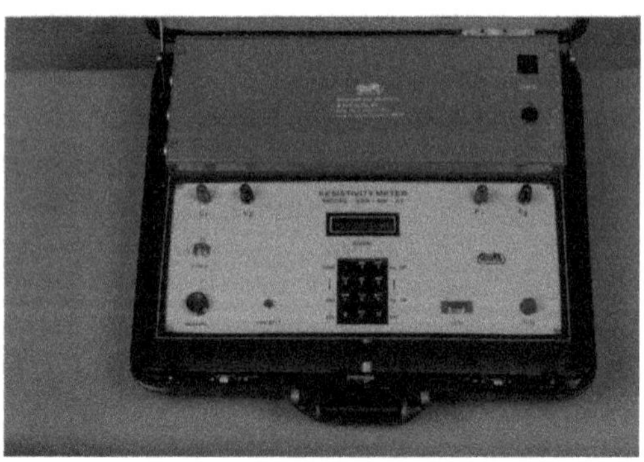

Fig. 6.9 Microprocessor – based signal stacking resistivity unit.

(Courtesy: Integrated Geo Instruments & services, Hyderabad)

3. The electrodes are kept 1m apart from each other (a = 1m) and the battery source is switched on (current is passed into the ground). The potential difference ΔV induced with the current I (noted on the current meter) is recorded.

4. The spacing, a, in the profile is changed to the next value (say 2 m). The electrodes are spread symmetrical to the centre of the profile. With current passed into the ground, the values of I and ΔV are recorded for this electrode spacing.

5. The procedure is reported for different electrode spacings and corresponding values of 'I' and 'ΔV' are recorded every time.
6. The field record is tabulated as

A (m)	I (ma)	ΔV (mv)	Resistivity
-	-	-	-
-	-	-	-

{Resistivity, ρ_a (ohm. m) = $2\,\Pi\,(\Delta V/I).\,a$ for Wenner's spacing}

7. Values of resistivity (ρ_a) are plotted versus "a" on a log-log plot.

Typical plots for 2-layer and 3-layer cases are indicated in Fig 6.10.

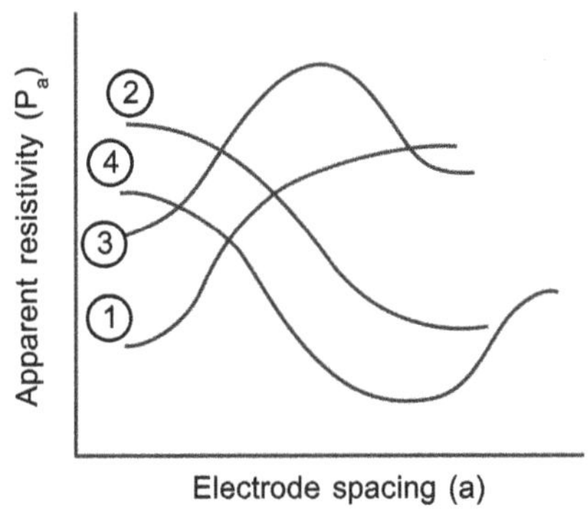

Fig. 6.10 Typical resistivity curves for two and three layers. (Curves 1,2 for 2 layers; 3,4 for 3 layers)

As already indicated, the resistivity values, thus computed, approximately correspond to the depth values equivalent to the electrode spacings , using Wenner's spacing. However, if exact depth calculations are to be made with this field data, master curves and auxiliary curves are to be used and the prescribed procedure is to be followed for the estimates of true resistivity and exact depth values. For 2-layer situations, the master curves corresponding to different resistivity ratios (μ) are of two categories - ascending (for various ratios of μ with ρ_2 greater than ρ_1) and descending (corresponding to ratios of μ with ρ_2 less than ρ_1). The auxiliary curves are available for different combinations of resistivity ratios (μ) and corresponding ratios (ν) of layer thickness values. The field curve is matched with the appropriate master curves / auxiliary curves. While matching is done, it is imperative that the field curves should be plotted to the same scale on a log-log plot (of the same modulus as the master curves). The master curve–auxiliary curve sets for Wenner's configuration are different from those for Schlumberger's configuration. These master curves are available for 2-layer and three-layer situations. For the latter, the curves are designated as follows :

H type : $\rho_1 > \rho_2 < \rho_3$

K type : $\rho_1 < \rho_2 > \rho_3$

A type : $\rho_1 < \rho_2 < \rho_3$

Q type : $\rho_1 > \rho_2 > \rho_3$

For situations with more than three layers, combimations of these types corresponding to the sequence of resistivity values are used to describe the individual cases (Bhattacharya and Patra, 1968).

A typical field curve for a multi-layered situation in Gangetic Plains is indicated in Fig. 6.11 for illustrating the curve-matching procedure. The field curve is drawn on a transparent sheet (with the same log-log scale as the master curves). The matching of the field curve is done part by part (ascending and descending) with master curves of that type (ascending or descending). During matching the corresponding X and Y-axes of the field and master curves are kept parallel to each other.

Layer	r (am)	h (m)	H(m)	Lithology
1	110	3.8	3.8	Artificial fill
2	71.5	18.62	22.42	Coarse sand with fresh water
3	14.4	25.56	47.98	Silly clay
4	16.5	97.75	145.73	Silly clay
5	37.0	-	-	Fine sand

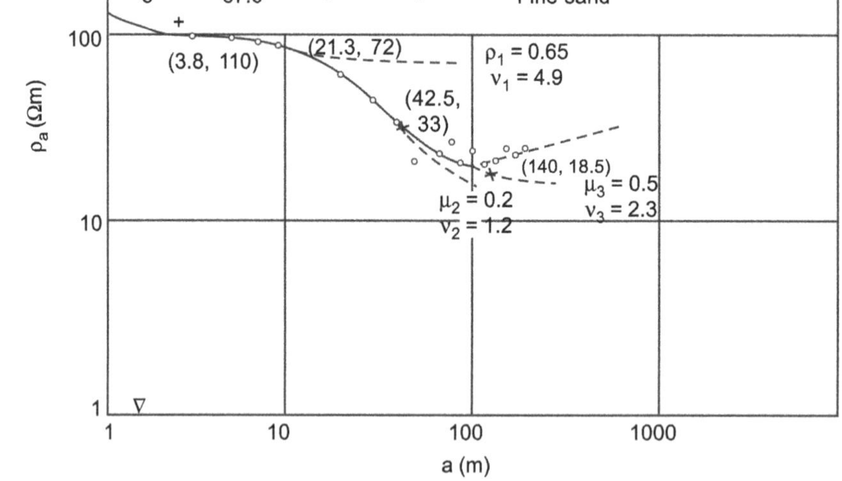

Fig. 6.11 Resistivity curve for alluvial sequence in the Unnao area (U.P.). (V.N. Bajpai, 1989)

The first part of the field curve matches with a descending type curve corresponding to the resistivity ratio (μ) of 0.65. In the matched position, the coordinate point of the master curve is traced on the master curve and indicated with the + symbol. The X and Y coordinates correspond to the values of 110 and 3.8 on the field curve and these are also indicated below the + point on the field graph. These directly correspond to the resistivity and thickness values of the first layer. Since the μ (ρ_2/ρ_1) value is 0.65 , the ρ_2 value is estimated as 71.5 based on the ρ_1 value of 110 (already estimated). The procedure is repeated for the subsequent parts of the field curve. It may be noted that only for the first layer the coordinates of the match point (+) are directly taken as the resistivity and thickness values of the layer. For subsequent layers the corresponding match points have to be used in the calculations for getting the resistivity and thickness estimates. The true thickness values of all these layers are obtained by matching them with the auxiliary curves and noting the ν (ρ_2/ρ_1) values. For each of the parts of the field curve (Fig. 6.11), the μ and ν values are noted on the extended parts (dotted parts) of the matched master curve traces on the field curve.

The respective true resistivity and true thickness values for different layers, as interpreted from the resistivity curve, in this example are indicated in Fig. 6.11. These resistivity values are interpreted in terms of the lithology of the formations. For this, a broad idea of the type of formations in the site is essential.

Self- potential method

This method involves measuring the potential between the potential electrodes for different electrode spacing ('a') without any current input into the ground. This means the potential measured is the natural potential existing in the ground all the time and is not the induced potential as in the previous method. Natural potentials of considerable value exist in the ground in formations like clay, shale or any sulphide- bearing ores.

Electrical well logging

Electrode assembly is lowered in a drill hole and at regular intervals of depth, the readings are taken for material across on the wall of the well. Resistivity is computed if the field is an induced field. Just like the surface measurements as discussed earlier, the self-potential can also be measured without passing any current through the current electrodes.

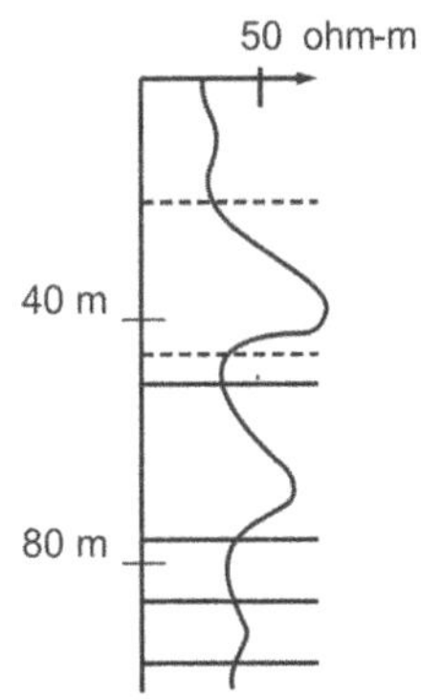

Fig. 6.12 Typical resistivity log.

In Fig. 6.12, typical resistivity is indicated. Similar to resistivity logging, S.P. logging can also be carried out. The procedure involves a continuous measuring of S.P. values with depth in a borehole. Both resistivity as well as the S.P. logs are not only useful in obtaining information on the nature and thickness of the formations, but also in establishing the productive aquifer horizons and the quality of the water contained in them. Details are indicated in chapter 7.

Drilling

Drilling and boring are undertaken for several purposes during the site exploration for civil engineering projects. During the preliminary investigations, bore holes are drilled to obtain information on the formation variations to depth. Delineation of bedrock depth is also carried out. This will provide information regarding the thickness of overburden in the area. Bore holes are also utilized for conducting tests like permeability and geophysical subsurface logging. Installation of piezometers in boreholes is done to monitor pressure changes. Sample collection from boreholes is one of the important components of site evaluation. Depending on the type of drilling, samples are obtained either as wash samples or as core samples.

Auger boring

This method is adopted for very shallow depths normally not exceeding 10 to 15 m. As the depth increases, the procedure is slow and inconvenient. The simplest unit used is the hand auger, which is rotated into the ground. It has a helical bit to enable the holding of sample when boring is advanced. Bucket type hand augers are very useful for collection of sample with least mixing.

Instead of operating manually, boring is also done with power-operated augers mounted on wheels. The material bored advances through the helical auger spiral or alternately collected at periodical intervals by lifting the unit.

Auger boring is undertaken in soils and semi-consolidated rock formations and is not suitable in the situations where the soil has gravel or cobbles. In the case of loose unconsolidated material, casing is essential to hold the material from falling as the boring advances.

Cable tool percussion drilling

Drilling is accomplished through impact and cutting actions of the bit. The tools are suspended on a cable with a facility to alternately raise and drop the drilling tool. The impact is accomplished through a weight (known as hammer), which is either on the drill rod above the ground surface or located on the drill bit, thus advancing down. In the latter case, it is called down-the-hole hammer drilling. The drill rod is operated through an air compressor. The drill mechanism enables the bit to rotate in between the hammer blow periods (Fig. 6.13). Air or water circulated through the drill pipe facilitates lifting up of the drilled sample. Thus the material is mixed and not very representative of the actual ground situation. This method is used in rock formations, which can easily be broken through impact and drilled easily.

Fig. 6.13 Down-the-hole hammer drilling unit.

In India, down- the- hole hammer-tool drilling methods are popular in the basaltic terrain in Maharashtra and hard–rock areas in Karnataka and Tamil Nadu.

Rotary drilling

Drilling is accomplished through rapid rotation and advancing of a drill bit attached to a column of drill pipe. The drill pipe is rotated by a motor at speeds varying up to about 2000 rpm. The fluid circulated into the hole brings out the drilled material. Entry of the fluid into the hole is done either through the hollow pipe or directly through the annular space between the drill rod and the wall of the hole. The former is adopted in the direct rotary drilling while the latter procedure is involved in the reverse rotary drilling

In the direct rotary drilling (Fig. 6.15), the drilling fluid sent through the pipe issues through the openings of the drill bit and mixes with the drilled sample. The water pressure facilitates continuous lifting up of the drilled material from the bottom of the hole to the surface through the annular space between the rod and the hole. In this process the material gets mixed up with loosened and dislodged material at different depths from the surface of the drill hole. There is thus a possibility of the material obtained from time to time of not being representative of the actual sample at any depth.

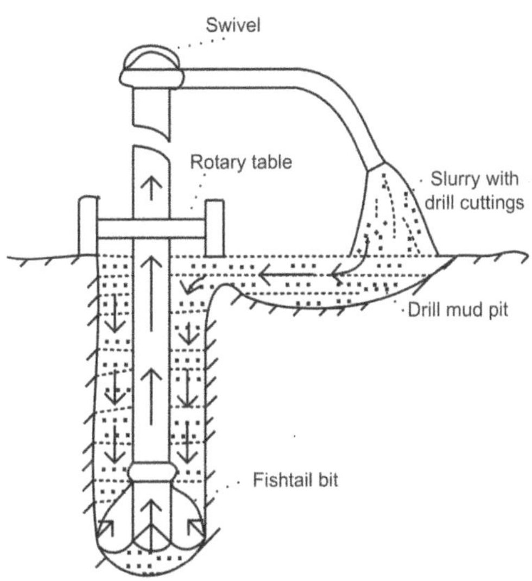

Fig. 6.14 Schematic diagram of reverse rotary drilling operations.

In the reverse rotary procedure (Fig. 6.14), the drilling fluid entering the hole directly (and not through the hollow drill pipe as in the previous case), lifts the material and carries it through the hollow drill pipe. Thus, the possibility of the sample getting mixed up with other material from different levels of the hole is avoided.

The drilling fluid consistency is important for lifting and carrying the drilled material. If the material has gravel or coarse-grained concretions as in the case of Gangetic alluvial material, the drill fluid consistency is achieved by mixing bentonite (approximately around 8 to 12 kg per 150 liters of water) within the circulating water pit on the surface (Fig. 6.14).

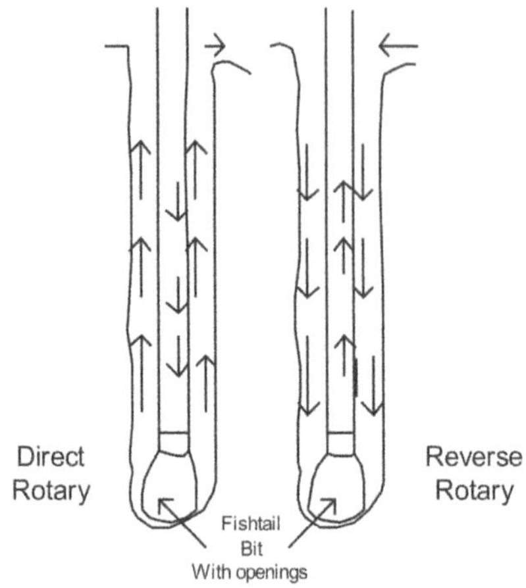

Fig. 6.15 Drilling fluid circulation in rotary drilling.

The drill bits for rotary drilling are of hardened steel or of tungsten carbide and are of different shapes. In the case of fishtail bits, the cutting face of the bit has bevelled radiating blades, in some cases with a slight curvature. Since these edges resemble fish tail, it is accordingly named. In the case of roller rock bits, the cutting face is composed of toothed rollers (Fig. 6.16).

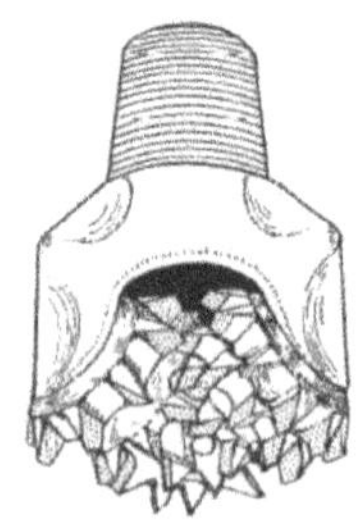

Fig. 6.16 Typical roller rock bit.

The individual rollers roll on the rock as the bit rotates. The roller teeth are made of steel or hard alloy. The star bits have wedge-shaped edges in the form of a star. If the bit has cutting face made up of rivet-shaped sharp protrusions, it is known as a button bit.

The rotary type of drilling, especially the reverse rotary procedure, is very commonly used for drilling the unconsolidated to semi-consolidated alluvial formations in the Gangetic Plains of U.P. and Bihar. This procedure is also very suitable for drilling soft rocks.

Diamond drilling

This is essentially rotary drilling, but the cutting face of the bit has industrial diamond or bortz. The bits are of two varieties.

- Core type bit
- Plug type bit

The core type (Fig. 6.17) is of hollow cylindrical shape with diamonds all along its edge. When the bit rotates, it scoops the material .As the drilling continues, the material is cored into a cylindrical shape. The core is thus issued into a barrel. In the non-core type and thus the sample is in the form of cuttings. The drill cores obtained depend on the diameter of the bits used. The sizes are classified into four categories (Table 6.4).

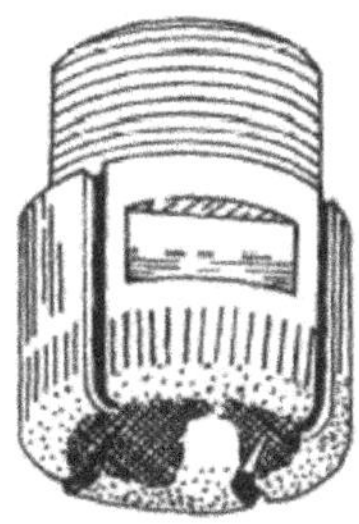

Fig. 6.17 Core type diamond drill bit.

Table 6.4 Diamond drill core sizes

Category	Diameter of the hole (mm)	Diameter of the core (mm)
EX	38.0	22.2 (7/8")
AX	46.5	28.6 (1-1/8")
BX	60.5	41.2 (1-5/8")
NX	76.4	53.9 (2-1/8")

For drilling to greater depths, in the initial stages NX bit is used up to about 15m, followed by BX (up to around 50 m), AX up to 80 m and finally EX beyond this depth.

Diamond core drilling is commonly undertaken in the site investigations for dams and tunnels. Samples in the form of cores are extremely important. Rock quality designation (RQD) is based on core recovery. Laboratory tests on rock cores enable engineering classification of the rock formations and also the designing of tunnel supports. Rock cores being the representative samples of formations at different depths, subsurface panel diagrams constructed on the basis of drill hole data provide an excellent subsurface picture regarding the geological and structural status of the subsurface

Types of drilling commonly adopted in different formations are indicated in Table 6.5.

Table 6.5 Suitability of drilling methods in different formations

Formation	Drilling
Soft and semi-hard	Light percussion type
Formations with rock pieces /boulders	Rotary cum percussion type
Unconsolidated and semi-consolidated	Direct rotary type
Alluvium (clay / sand / gravel)	Reverse rotary type
Hard rocks	Down the hole hammer drills
Hard rocks (for core samples)	Rotary core type

Recording of drill hole data : During the drilling operations, the sample is collected at regular intervals of depth. In the case of percussion or rotary types, the samples are in the form of wash samples. In the core drilling, the sample is obtained as core. The recovery of core in term of the length of continuous pieces depends on the frequency and spacing of joints in the rock. The drill hole samples are kept in a sample rack with slots and the depth is mentioned on each of the racks. Once the drilling operations are completed, the samples are classified geologically and strata chart is prepared. A strata chart indicates the formations in sequence logged to scale and respective thickness values recorded against the formations.

Drilling need not always be always in the vertical direction. In the case of steeply dipping formations, the drill holes are placed at right angles to the dip direction (Fig. 6.18).

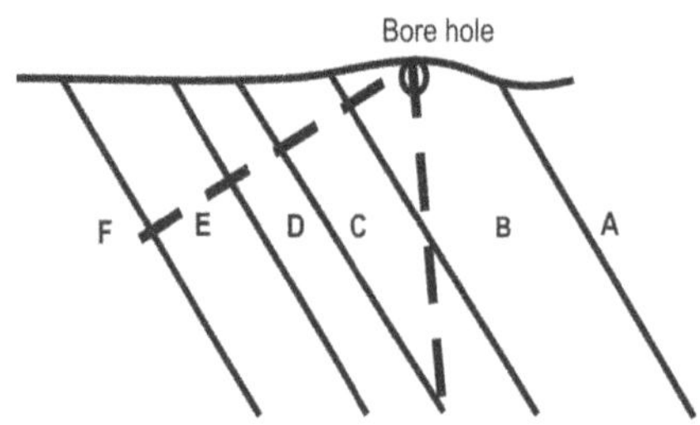

Fig. 6.18 Inclined and vertical bore holes in a dipping sequence.

This would enable intercepting of maximum lithological variations in a drill hole of chosen depth (Fig. 6.18). Incidentally, inclined drill holes facilitate monitoring of shear zones and faults.

Ground Improvement Methods

Grouting

Grouting involves injection of an aqueous suspension of solids (commonly cement-based) in the form of slurry or a uniform chemical mixture under pressure into the ground through drill holes or pipes to seal all weak zones in the formations. The injected grout hardens with time.

Grouting is under taken to :

1. Improve the strength and bearing capacity of the formation
2. Block seepage by reducing the permeability of rock and soil formations
3. Completely fill underground openings such as solution cavities
4. Fill shear zones with crushed material making the entire material as a single solid entity effectively tied with the surrounding solid rock.
5. Raise and re-level structures and foundation elements that have settled uniformly or otherwise
6. Augment pile capacity and pile repair

Types of Grouts

1. *Suspension grouts*

These grouts involve materials such as portland cement in the form of slurry. As the name implies, the grout material is in the form of suspension. A higher water to cement ratio is adopted to facilitate smooth injection of the grout slurry into the weak zones. The initial ratio is varied during the later stage of operation, once the initial grout has completely penetrated the formation voids all around in a drill hole.

2. *Solution grouts*

These are a non-particulate grouts in that the material is present as a solution. The grout materials consist of two liquids, which form into a gel on mixing and harden with time (setting period). The setting varies with the type of material used. For example, silica gel is prepared using sodium silicate and calcium chloride.

Solution grouts are employed in one of the following two ways :

One shot grout : Mixture prepared initially and used as the grout

Two shot grout : The two chemicals are injected successively into the hole to react and form the gel.

3. *Organic resins including epoxy resins*

Most of the grouts are air-entrained. In selected cases, foam grouts are also employed. Foam grouts, being of gaseous emulsion of an ordinary grout, involve relatively lesser quantity of the grout material. Suitable foaming agent is added to create foam.

Choice of grouts depends on the nature of materials to be grouted. (Table 6.6).

Table 6.6 Principle grouts employed (Modified from Cambefort, 1977)

Grout Type	Cement	Bentonite and Cement	Deflocculated bentonite cement org.	Chemical Products Na-silicate Org. Resins	Aerated Grouts
State	(———	suspensions———	———)		
	unstable	——— stable ———		——— liquids———	aerated emulsions
Range	fissures	———	sand and gravel ———	——— cavities	high water flows
		$k>5x10^{-4}$ m/s	$k>10^{-4}$ m/s $k>10^{-4}$ m/s	$k>10^{-4}$ m/s $k>10^{-6}$ m/s	

Grouting Procedures

1. *Penetration grouting*

A. *Intrusion grouting :* Cementitious grouts are injected through drill holes to fill cracks and joints.

B. *Permeation grouting :* Thin grout, commonly chemical grouts, injected into material to fill easily the voids. This is also adopted during tunneling operations to make the formation hard enough for tunnel excavations.

2. ***Compaction grouting or displacement grouting***

   Thick stiff grout is injected under high pressure through a pipe. This method is often adopted to repair structures, which have undergone excessive settlement.

   The grout mass injected through a grout pipe set in a drill hole, forms a bulb around the place where it is injected. The compaction grouting procedure is carried out in different steps. At depth, it is injected first and the grout pipe is lifted up to around 1 m and the grout is again injected. This process is repeated by pulling out the grout pipe in stages until the entire formation layer is grouted.

3. ***Jet grouting***

   Jet grouting is a special method developed in Japan. A special pipe equipped with horizontal jets is used to forcibly inject the grout material laterally. The grout mixes thoroughly with the soil resulting in a stronger and impermeable grouted mass of the soil. This method has a wide range of uses including construction of cutoff walls, deep foundation treatments and in underpinning operations.

Typical grouting applications

In dam construction, grouting is carried out through grout holes prior to construction as in the case of earth and masonry dams. In the case of concrete gravity and thick arch dams, grouting is undertaken through galleries and grout caps during the construction of the dam. For thin arch dams, the grouting is done through grout caps on the upstream heel.

The details of grouting are as follows :

1. Consolidation grouting (also known as blanket grouting)

   Grouting in this case is to be carried out and completed prior to the construction of the structure. The grout holes are placed in a grid pattern at specific spacing and direction so as to cover the desired portion completely.

2. Curtain grouting from a gallery or a grout cap (Fig 6.19)

   For earth and rock-fill dams, the grout curtain is placed along the axis of the dam prior to the construction of the structure. However, the grout procedure is adopted, after the construction, through grout caps at the upstream heel of the dam.

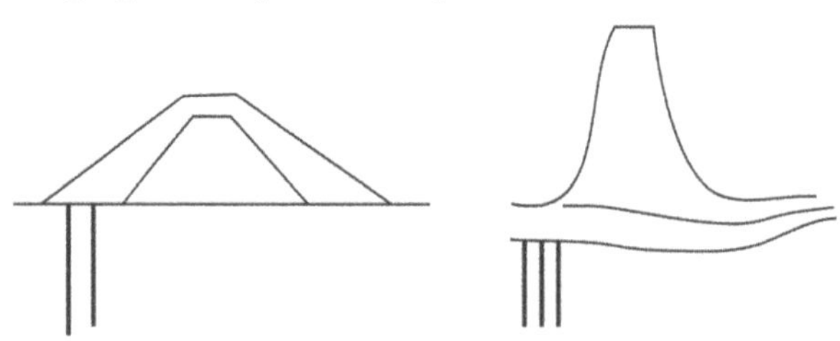

Fig. 6.19 Grout curtains in dam foundations.

   Curtain grouting is very effective if undertaken after the construction just around the period of application of load on the foundation. Normally, high pressure is adopted for the grouting to overcome resistance from the adjacent material. The depth of the grout curtain is dependent on the local geological conditions. However, commonly the vertical depth is chosen to be as one-third of the dam height at the grout hole location plus an additional depth of 20 m.

3. Off-pattern grouting in special situations

   Commonly during the design of the structure, the grout scheme is prepared in detail for anticipated geological situations. The hole lay out; spacing, directions respective types and depths including the amount of grout to be placed are all planned and estimated. This is included under the planned pattern grouting. However, as the excavation proceeds and construction progresses, unanticipated geological situations may be encountered for which grouting is needed. The off-pattern grouting belongs to this category.

   The grouting undertaken in tunneling, excavation zones and for filling solution cavities in limestones is indicated in Figs. 6.20 to 6.23.

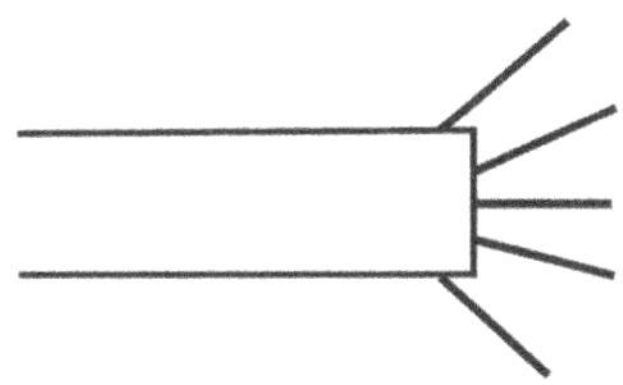

Fig. 6.20 Grouting in tunnel advance.

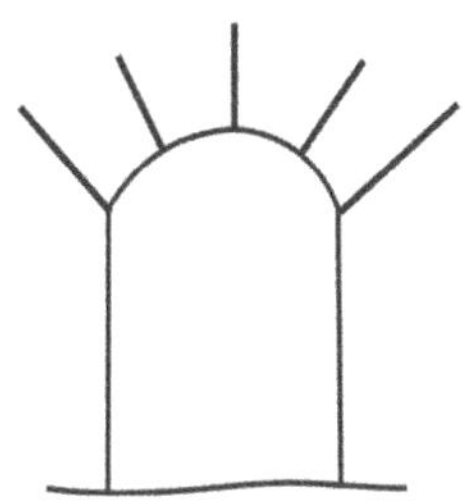

Fig. 6.21 Grouting on tunnel linings.

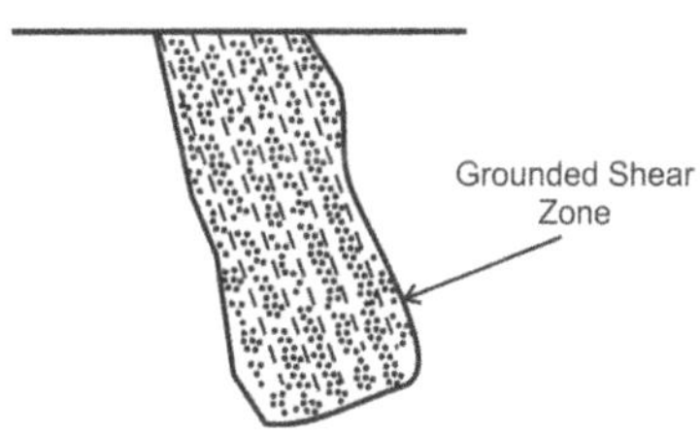

Fig. 6.22 Grout-fill in excavated shear zones.

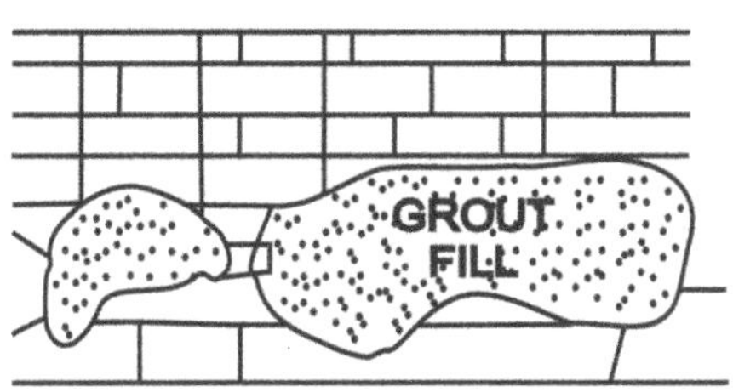

Fig. 6.23 Grout filling in solution cavities in limestone.

For backfilling : Excavation of weak zones is done normally up to a depth specified by the height of the structure: (The units as specified originally are retained here)

$$D = 0.002\ bH + 5 \text{ (for } H < 150 \text{ feet)}$$

$$D = 0.3\ bH + 5 \text{ (for } H > 150 \text{ feet)}$$

where H = the height (above foundation level) of the dam (in feet)

B = width of the weak zone (in feet)

D = depth of the weak zone (in feet)

In the case of clay gouge, "d" should not be less than 0.1H.

In the case of Bhakra dam, the depth (D) used, as per the formula of the U.S. Bureau of Reclamation, is :

$$D = 0.002\ bH + 10 \text{ (all the units are in feet)}$$

Grout Consistency and Pressures

The consistency depends on the type of void spaces and the nature (like particle size) of the grout suspension. In the engineering terminology, the grout suspension is described as "thin and slimy" or "thick and stiff". Thin suspensions have considerable quantity of water (up to about 8 times in volume compared to the volume of the grout material) while in the case of thick grouts; water-grout material ratio normally corresponds to 1:1.

Shotcrete

Shotcrete is mortar or concrete pneumatically sprayed at high velocity through a hose. The difference between concrete and shotcrete is that shotcreting involves both placing and compacting together while in case of concrete, it is first placed and compacted later. Shotcrete achieves an excellent bond with the surface it is applied on.

The process can be a dry process or a wet process (Hoek, 2000). The dry process is also known as guniting. In the dry process, dry or semi-wet material is conveyed to a nozzle through a hose (of about 5 to 6 cm in diameter) in the equipment at high velocity pneumatically. Water is added at the nozzle end (Fig. 6.24). Compaction on the surface is also achieved while it is delivered onto a surface because of the high velocity impact involved. The process is easy to operate and has the advantage of controlling the material at any state. However, enough compressed air pressure is needed because of the relative dryness of the material. On the other hand, in the case of wet process, the materials are premixed with desired water content and pumped to the nozzle where the compressed air is introduced (Fig. 6.25).

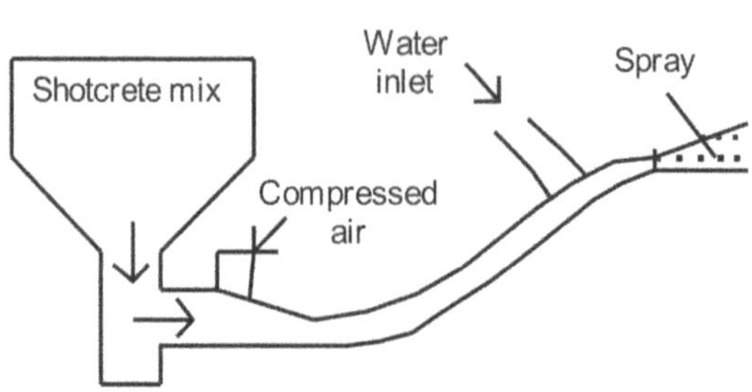

Fig. 6.24 Schematic sketch of the equipment for dry process.

The wet mix is delivered at high velocity onto the surface. The aggregates (around 0.8 cm in size) are limited to 20% of the mix.

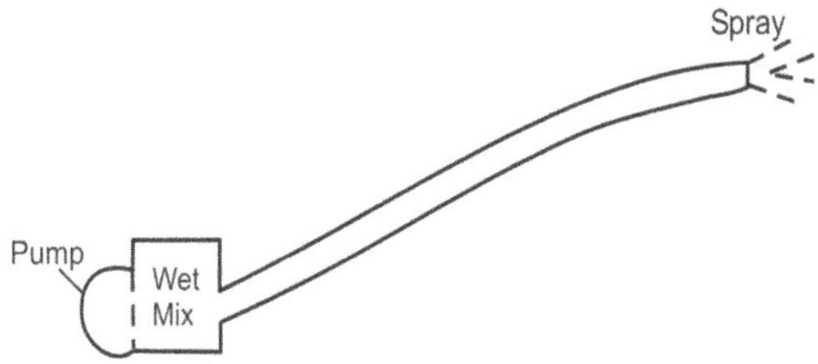

Fig. 6.25 Schematic sketch of the equipment for wet process.

To increase the strength and for improved bonding, silica fume (up to about 10% by weight) is added to the mix. The mix in such a case is sticky which enables easy placement of a relatively larger mass onto the surface. Shotcrete is ordinarily brittle on hardening. However, addition of steel fibres to the shotcrete mix (known as steel fibre-reinforced shotcrete), imparts greater ductility to the material, which can contain non-elastic deformation to a greater extent. A typical mix design of this shotcrete has 18-19% cement, 1.7-2.2% silica fume additive, 68-75% blended aggregate, 2.7% steel fibres and 0.6% accelerator. In the case of dry mix, controlled addition of water at the nozzle is carried out while in the wet mix, about 7.5% water is added. Quite often, wire mesh is used for achieving enough stiffness to this grout. The grout strengthens rapidly attaining compressive strength upto 4.2 MN/m^2 at the end of 28 days and thickness of 200-300mm can be achieved. During tunneling operations, shotcreting preceeds rockbolting as an additional reinforcement. The grouting pressure is to be carefully controlled such that it will not shatter the formation. On the other hand, the pressure should enable not only the free entry of the grout into the void spaces but also its free circulation to different areas in the formation to be grouted. In the case of the reservoirs already filled, the grout pressure for curtain grouting is maintained as 1.5 to 2 times of the estimated hydrostatic pressure of the reservoir water at the grouted depths.

Rock bolts

These are steel bolts designed for holding weak formations together. The bolts are driven into the formations without causing any disturbance or dislocation. These are used in tunneling for anchoring the tunnel walls to solid rock. In potential landslide regions, rock bolts are used for rendering the slopes stable by reinforcing the slide area.

The rock bolts are of different forms :

1. *Slotted bolts*

Individual bolts are about 6 feet (around 1.82 m) in length with a diameter of 1" (2.54 cm) with threads on one end and slotted and bevelled on the other. When the bolt is driven into the formation through a hole slightly larger, it wedges itself with widening of both sides of the slot with the advance of the bolt into competent strata. Once the bolt is driven, a washer plate is placed on the threaded area and tightened with a nut. A typical slotted bolt is shown in Fig. 6.26.

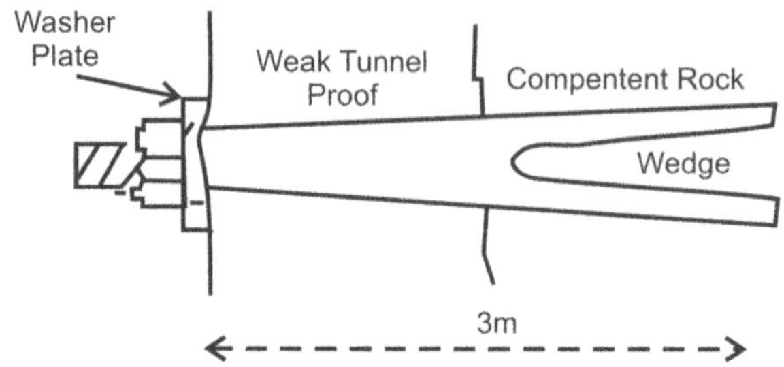

Fig. 6.26 Slotted bolt.

2. *Perfobolts*

These are used for anchoring the weak rock with a solid competent zone away from the same. The perfobolt system consists of a series of prestressed high-tension wires of about 7 mm diameter. From the main tunnel, these prestressed anchor wires are connected to an auxiliary tunnel located in the competent portion of the rock (Fig. 6.27). Using stiffeners on both ends of the system, these wires are tightened with the help of locking nuts. In the main tunnel, a thick shotcrete pad is provided on the hole (at the surface) as a strong supporting base.

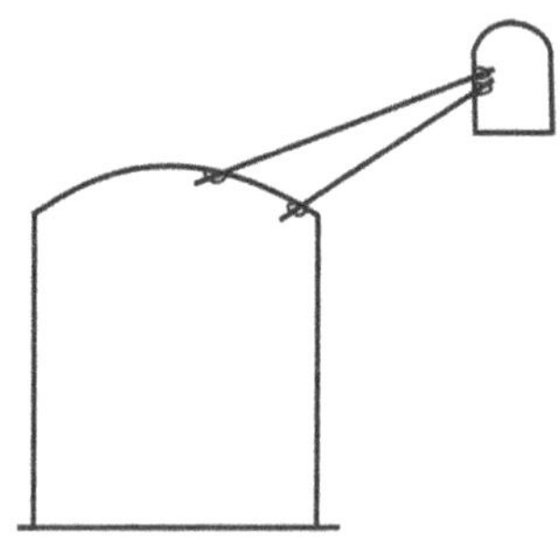

Fig. 6.27 Pre-stressed cable anchor.

This type of bolting is adopted in the soft Siwalik formations in which an underground power house was situated at Chibbro in Dehradun district of Uttaranchal. Details regarding the support system for this underground power house cavity are presented in chapter 9.

Heavy Tamping (Dynamic Consolidation)

Heavy tamping is resorted for achieving densification and improving the mechanical properties of soils. On the surface, dynamic compaction is carried out by systematic dropping of heavy weight (around 10-15 tonnes) from a height of about 20m with the help of a crane. Each impact made is backfilled and tamping is continued. This procedure is repeated at regular intervals at the same location. The locations are chosen on a grid pattern. The depth of compaction is a function of the weight and the height of its drop. Once the main compaction is achieved, final phase involves dropping the weight from a lesser height (around 4 to 5 m).

For compacting soils at depth, vibro-compaction (vibro-densification) technique is adopted. The procedure involves inserting a vibratory probe which produces vibration. The penetration of the probe is accomplished by water jetting at the tip of the probe. This method is also known as vibro-flotation. The soil improves its density through rearrangement of its loose granular particles. Vibro-compaction helps in improving the bearing capacity and shear strength. Soil settlement is also reduced. In quake-prone areas, this method helps in increasing the resistance to liquifaction.

Preloading

Soft marine soils are improved by preloading or precompression. The ground is loaded with earthfill of sufficient height to achieve the required amount of settlement. Once the settlement is achieved, the earthfill is removed and the proposed structure built. Thus the structure experiences only a small fraction of the settlement as the major component of settlement has taken place during loading with earthfill and only that part of settlement due to unloading and reloading occurs due to structural loading. The main limitation of preloading is that very large times are required for consolidation to occur in fine-grained soils such as marine clays.

Preloading vertical drains (sand or strip) is the most preferred alternative. The drains installed in a triangular or square grid pattern and at relatively closer spacing reduce the length of the drainage path and reorient the direction of flow from vertical to the horizontal and thus achieve 90 to 95% degree of consolidation in relatively short period (6 to 9 months) instead of several years. The strip drains, 100 mm wide and 4 mm thick, are made of non-woven filter with a core inside. They are installed rapidly in soft clays even up to depths as great as 20 m.

Soil stabilization

This is commonly adopted in the case of expansive soils to improve strength as also to reduce swelling and shrinkage. In some instances, it is also resorted for stabilizing loose alluvial soils. The process involves mixing admixtures into the soil and compacting the soil at optimum moisture conditions. If the additive is only to improve the density of the soil and does not react with the soil, the process is known as mechanical stabilization.

If the admixtures chemically interact with the soil ingredients (the granular and colloidal fractions), the process is designated as chemical stabilization. The main intention in this case is to achieve artificial cementation in the soil mass, thus improving the strength and reducing swelling, shrinkage and permeability of the soil. The mixing has to be effective and the process of hardening is time-sensitive. During ageing period, pozollanic reactions in the soil-additive system result in the formation of strength-imparting reaction products. Additives commonly added include portland cement, flyash, lime, blast furnace slag and rice-husk ash. Lime, being of low-cost and reactive, is the most common additive. Lime raises the pH of the soil-additive system. Under the enhanced pH of the system, the soil minerals decompose resulting in the dissolution of silica. In the hydrated soil-lime system, the Ca of lime reacts with the silica and alumina to form calcium silicate hydrate and calcium aluminate hydrate phases as products on ageing with time. Just as in the case of cement on hydration, in the lime-stabilized soil the calcium silicate hydrate that forms with long-term ageing is the main strength-imparting pozzolanic product. In the hydrated cement it is tobermorite (a particular calcium silicate hydrate phase). However, compared to the reaction in the hydrated cement, the rate of reaction in soil-lime system is relatively very slow and the quantity of the product is also less.

Black cotton soils cover an extensive part of Maharashtra and the neighbouring areas in M.P., Andhra Pradesh and Karnataka. These soils are expansive in nature. Soil stabilization offers an excellent scope for ground improvement (not reinforcement) in activities such as road construction.

Electro-osmosis is used in stabilizing soils of low permeability. Dewatering is done in this process by planting two electrodes into the ground to a depth of about 1.5 to 2m and energizing the ground with a D.C. battery source. The flow of water takes place from positive (anode) to negative (cathode) electrode points. The electrode connected to the negative point is in the form of a slotted pipe to function as a collection point for the drained water. A series of cathodes and anodes can be placed in an array to effectively drain the area.

Geosynthetics (Fig. 6.28)

These are synthetic permeable membranes increasingly in use over the last few years in civil engineering practice. They possess several properties including high strength, flexibility, resistance to fungal attack, high abrasion resistance, good performance under repetitive loading and chemical stability in water. They are used in ground reinforcement, slope stability, road construction, protection from drainage, separation and erosion control. Essentially, the material used can be classified into two main categories:

1. ***Biodegradable geosynthetics :*** These are natural products made out of material like jute and coir. They are used for temporary purposes like in makeshift roads and they degrade in one or two years, by the time the road is decommissioned. They have very high moisture absorption capacity (400-500%), flexibility and water drainage properties. Available in many forms as open mesh, woven forms or unwoven fabric, these are of low-cost.

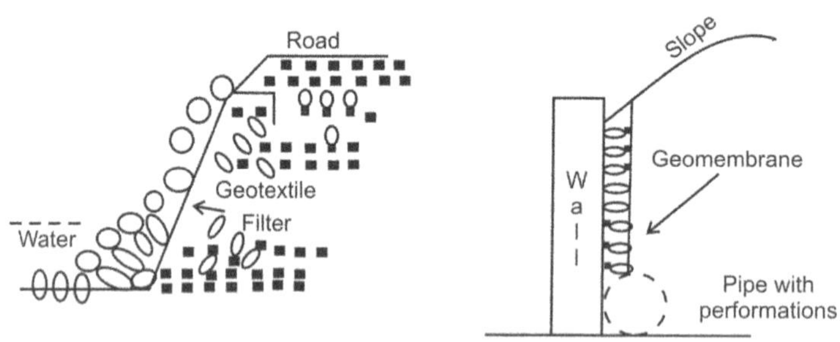

Fig. 6.28 Uses of geomembranes for slope stability.

2. ***Non-degradable geosynthetics :*** These are of durable material. These are also available as woven and non-woven forms.

   Geotextiles are being extensively used in the Golden Quadrilateral and North-South and East-West corridor projects of National Highway Authority of India (NHAI) as separators between the subgrade and the base course. They enhance the life of the pavements by several years. Geogrids are the principal component in reinforced earth structures, which are being invariably used in all flyover approaches in several cities. Geomembranes form the main barrier in containing and preventing hazardous leaches reaching groundwater in all municipal solid waste landfills.

7

Hydrogeology

Hydrogeology deals with occurrence, distribution, storage and movement of groundwater in the subsurface. Groundwater is an important source of water supply and is used in conjunction with the water from surface sources such as streams, lakes and rivers. Unlike surface water, groundwater needs very little treatment. It is relatively free from pollution and the dissolved solids acquired from the host rocks normally control the quality. For example, groundwater in a limestone terrain is rich in calcium and is typically hard water in its nature. Use of such type is not recommended in boilers due to the formation of scales on the boiler linings. Similarly, in the granites in several parts of the country flouride content in the groundwater is in excess of the permissible limits for its use for drinking purposes. The details regarding the flouride hazard are detailed in the chapter 10.

All water below the earth's surface is referred to as the groundwater or subsurface water. As the water from the surface percolates or infilters into the ground, its distribution and movement in the subsurface is guided by the porosity and permeability of the geological material (soils/rocks). Porosity or the pore space in these materials facilitates the storage while permeability contributes to its movement.

Ground water is one of the components of the hydrologic cycle in nature. Hydrologic cycle enables a clear understanding of the recharge, storage and movement of water in the subsurface.

Hydrologic cycle

Hydrologic cycle depicts the continuous recirculation of water in nature. The different paths involved have different time scales of movement. Three aspects are involved in the hydrologic cycle, namely transportation, storage and change of state. In any of the paths, some of these or all of them are involved. For example, water in the snow is in a solid state of existence, which on melting results into the liquid form. In evaporation, the gaseous state is operative. A change from one to the other of these stages occurs in the hydrologic cycle. Transportation of water is in the form of surface flow, evaporation, transpiration or groundwater flow.

Hydrologic cycle enables an understanding, development and management of water resources. In Fig. 7.1 various components of the hydrologic cycle are shown.

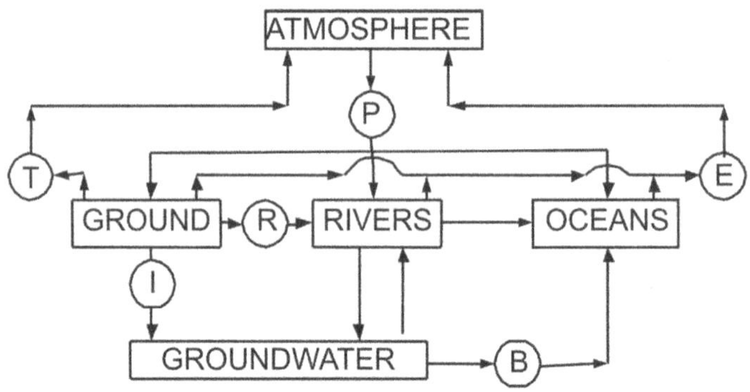

Fig. 7.1 Hydrologic cycle.

(**P** = precipitation, **R** = run-off, **I** = infiltration, **B** = base flow, **E** = evaporation, **T** = transpiration)

A convenient starting point to start from, in this continuous circulation of water in nature, is the precipitation of water from the atmosphere. Precipitation takes place in the form of rain, snow, hails or dew. Once the precipitation reaches the ground surface, the water channelizes into two parts –one on the ground as surface flow and the other part that infilters into the ground contributing to the groundwater component of the cycle. It may also be noted that the vegetation also intercepts a part of the rainfall before it reaches the ground surface and returned to the atmosphere as evapotranspiration. The surface flow, called the overland runoff, is through streams, lakes and rivers. Part of this surface flow is stored as surface storage at convenient locations such as lakes. The surface flow ultimately reaches the ocean. On the surface, loss of water is in the form of evaporation from the surface water bodies including the ocean and evapotranspiration from vegetation.

Contribution to the groundwater storage is not only directly from the precipitation on the ground but from various favourable recharge areas from the overland runoff. Just as there is movement of surface water in the form of runoff, movement of groundwater also takes place depending on its elevation (from a groundwater high to a groundwater low). Groundwater storage in the subsurface materials, as already indicated, is within the pore spaces and openings within the saturated zone. In India, about 90% of the precipitation for evapotranspiration (40%), runoff (40%), soil moisture (around 10%), and the rest 10% available as recharge to the ground water. However, these figures are an average and vary from place to place depending on the elevation, climatic condition and geology of a region.

All the inputs and outputs in a hydrologic cycle can be balanced in the system in terms of a hydrologic equation, popularly known also as the water-budget equation. Water budget equation can be written, in general form, as

$$\text{Mass Inflow} - \text{Mass Outflow} = \text{Mass Storage}$$

$$(\text{Precipitation})$$

$$[\text{Precipitation}] - \begin{bmatrix} \text{Surface runoff} \\ + \text{Evaporation} \\ + \text{Transpiration} \\ + \text{Net flow of groundwater} \\ \text{from catchment} \end{bmatrix} = [\text{Ground water storage}]$$

Thus, the water-budget equation for a catchment area is useful in the estimation of ground water storage if the other parameters for the area are known. Hence this equation is of paramount importance in groundwater modelling for a catchment area or a river basin. For modelling in the canal command areas in a catchment area, recharge from irrigation canals is also an input. Data regarding the rainfall, evaporation and transpiration are available from the meterological reports. Data regarding the overland flows (like the river discharge) are available in the hydrological records.

Underground water

Underground water occurs in different forms (Fig. 7.2). The portion below the water table is known as groundwater. Below the water table the entire zone is saturated. A capillary zone exists as a fringe above water table where the water is held by capillary action.

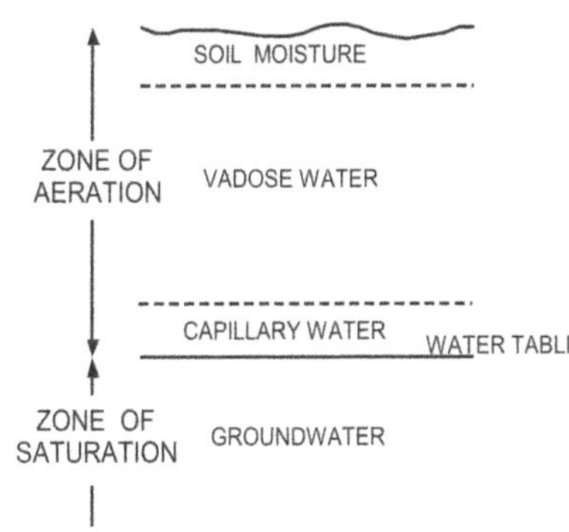

Fig. 7.2 Forms of subsurface water.

Water Table

Water table is the upper limit of the zone of saturation in the subsurface. Zone of saturation is the zone in which all the available pore spaces in a soil or rock are completely saturated with water. Water table fluctuates with the recharge inputs. In any area, the water table depths are different for the pre-monsoon and post-monsoon periods. During monsoon, the recharge of precipitation from the surface contributes to the ground water storage. Thus, the difference between these two depths of groundwater table provides an estimate of the additional volume of the saturated material due to the monsoon.

Water table contour maps

In groundwater investigations, the depth to the water table is measured at different places through wells or depth piezometers. Water table depths can also be estimated using electrical resistivity technique (details in the subsequent pages available under the groundwater exploration methods). Once the point values are obtained in the field for an area, these values are plotted at relevant locations on the map of the area. Then, a contour map is prepared by drawing contours of equal groundwater depth values. The contours are at chosen contour intervals similar to the contour maps prepared from the elevation data in surveying. The groundwater depth contour maps for both the pre-monsoon as well as the post-monsoon periods can be prepared for any region using the water table depth values as obtained in these two periods. These contour maps will be of help in locating new wells and also to understand the availability of groundwater after the monsoon period so that successful management can be planned for the supply of ground water till the succeeding monsoon period.

In addition to the contour maps for depth to ground water, contour maps can also be prepared from the groundwater table elevation data instead of the depth values of the water table. For this type of contour map, the individual point values of the water table depths are converted into the absolute elevation. For this purpose, the absolute elevation value of each of the data points is obtained from available sources like the Survey of India toposheets and the depth of water at each point is subtracted from its absolute elevation. This provides the absolute elevation (above mean sea level) value for the water table at each point. A contour map (Fig. 7.3) prepared on the basis of these reduced water table elevation data represents the absolute configuration of the water table as a planar surface. In this type of contour plotting, the effect of surface topography on the ground water elevation is removed. Widely spaced contours indicate higher permeability. In Fig. 7.3, area B has relatively higher permeability compared to area A.

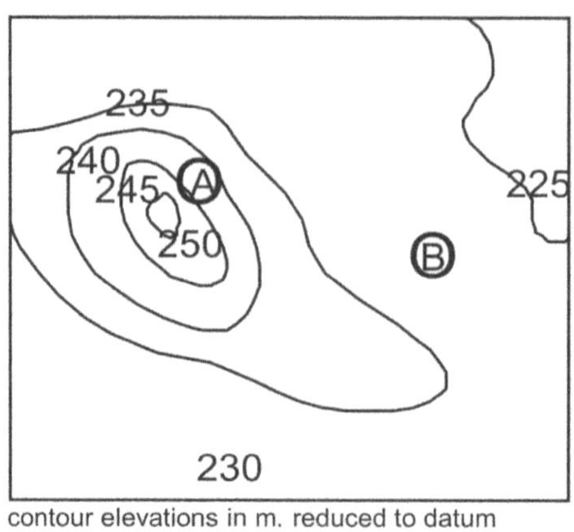

Fig. 7.3 Water table elevation contour map.

Uses of Water Table Contour Maps

1. Depth to the ground water table and monsoon inputs can be worked out for any location.
2. Areas of ground water-high (water table mounds) and ground water-low (water table depressions) can be located on the absolute elevation contour map for an area. The water table mounds denote that the water table in those particular locations is closer to the surface. This is due to the recharge in such locations from the surface to the water table. Hence, areas of recharge can be identified on a map for a region. These areas of recharge can be protected and improved for more effective recharge by providing adequate surface storages (like tanks or minor reservoirs) at such places to enable increase in the recharge.
3. Possible directions of groundwater flow can be drawn on the absolute elevation water table contour maps. The groundwater migrates from areas of higher elevation to areas of lower elevation.
4. Areas with high hydraulic conductivity of the geological material can be identified in contrast to the areas with low hydraulic conductivity material as the water table contours will be widely-spaced in areas with high hydraulic conductivity material. If a new well is to be located and if there is choice, areas widely-spaced contour maps are chosen as potential areas for good supply.
5. For the estimation of groundwater storage from the contour maps, the volume of the saturated zone is to be multiplied by the effective porosity thus obtaining the volume of the pore spaces saturated with ground water.

Classification of geological materials in hydrogeology

Soils and rocks are classified in terms of porosity and permeability to define their performance in ground water development. These two parameters are defined as follows :

Porosity (P) is the ratio of openings (voids) in the soil or a rock to the total volume (V_1) of the soil or rock, expressed as percentage.

Hence $P = \{(V_1 - V_2) / V_1\} \times 100$, where V_2 is the volume of solids.

Unconsolidated materials have better porosity

Porosity is classified as primary porosity and secondary porosity. Primary porosity includes openings (voids) developed at the time of formation of a rock or soil. For example vesicles in basalt, intergranular pore spaces in sandstone and intergranular pore space in soils belong to this category. Primary porosity is thus related to the origin of the geological formation.

However, if subsequent processes operative after the formation of the material develop openings in the same, these voids are designated as secondary porosity. Joints, fissures and solution cavities belong to this category. All these are openings formed in the material subsequent to its formation.

The porosity that we encounter in a geological material is thus a sum total of both the primary and also the secondary types.

Porosity, being a measure of the total pore space, aids in the groundwater storage in a soil or rock. Thus formations with good porosity have better storage of water.

Permeability

Permeability (k), also called hydraulic conductivity, is the ease with which water can flow through a rock or a soil. It is a measure of interconnected void spaces.

Hydraulic conductivity can be easily defined on the basis of Darcy's equation.

According to Darcy's law, Q = k. I. A

Q = quantity of water per unit of time

k = hydraulic conductivity

I = hydraulic gradient (dh/dl)

A = area of cross-section through which the flow takes place

By transposing the parameters, k = Q / (I.A)

Thus hydraulic conductivity is defined as the quantum of water that passes through an unit area of cross-section under unit hydraulic gradient. Hence, the unit of k is m/d {derived from m^3 / (day.m^2)}. Hydraulic conductivity has the same units as velocity.

Hydraulic conductivity may be same or different in different directions within the soil or a rock. If it is same in all the directions, the material is designated as isotropic in contrast to the anisotropic material with the flow differing in different directions. For example, stratified geological formations are anisotropic. The permeability in the horizontal direction in such a case is commonly more compared to the same in the vertical direction.

While a rock or a soil has isotropic or anisotropic flow at any particular point, the material need not necessarily have the same flow condition (isotropic or anisotropic) elsewhere within the same. If the flow is uniformly maintained as isotropic or anisotropic throughout the rock or soil at all the locations, it is considered as homogeneous. Otherwise it is called heterogeneous. So a rock or a soil can be homogeneous-isotropic/homogeneous-anisotropic/heterogeneous-isotropic/heterogeneous -anisotropic. Based on the porosity and permeability (hydraulic conductivity) characteristics, all geological materials are classified to denote their water-bearing and water-yielding properties. The details for the same are as follows:

Aquifer **:** If the rock or a soil has both porosity and permeability, it is designated as an aquifer. The material not only should store water but also should yield water when it is drained. In ground water investigations, the primary effort is to locate and delineate the aquifers. Unconsolidated or partly consolidated soil and porous sandstone are good aquifers.

Aquiclude **:** If the material has porosity but no permeability, it is called an aquiclude. It can store water but will not yield water as it is impervious and the flow of water doest not take place within the same. For example, clay is an aquiclude.

Aquitard **:** If an aquiclude has limited permeability, it is known as an aquitard. Clay with restricted amount of silt is an aquitard.

Aquifuge **:** If rock has neither porosity nor permeability, it is designated as an aquifuge. Massive compact granite without joints is a typical aquifuge.

Types of Aquifers

Aquifers are classified as

1. Unconfined or open aquifers
2. Confined aquifers
3. Leaky aquifers

Unconfined aquifer (Fig. 7.4)

This aquifer horizon is open to the atmosphere in the sense that it continues downwards directly from the ground. As such it operates under atmospheric pressure. Water is drained through gravity flow as the aquifer is drained during the withdrawal of water from a well located in the same. The water level in the aquifer is the water table, below which all the pore spaces are filled with water. Below the water table, the portion of the aquifer is known as the saturated zone of the aquifer. In groundwater investigations, in addition to locating an aquifer, it is necessary to delineate the water table since the productive zone of the aquifer (the zone supplying water) is from below the water table.

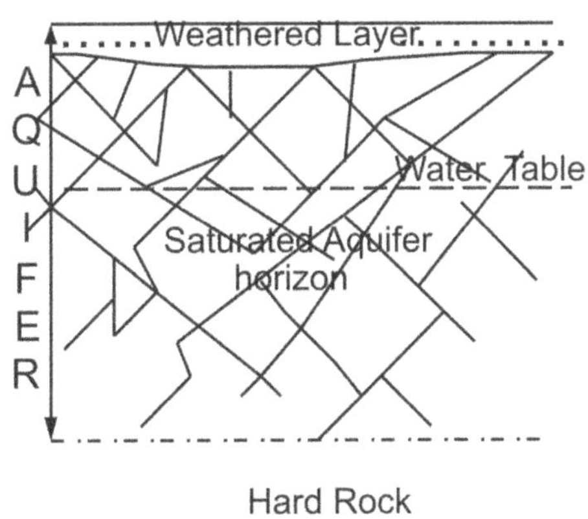

Fig. 7.4 Unconfined aquifer.

Confined aquifer (Fig. 7.5)

In this case, the aquifer is not directly open to the surface of the ground but is confined or sandwiched between two aquicludes. The confined aquifer has its recharge only from locations where it is exposed to the ground surface. So the groundwater storage is due to the infiltration at the recharge locations. The water in the confined aquifer does not operate under the atmospheric pressure as hydrostatic or the piezometric head controls it (Fig 7.5). Thus in a well located within the unconfined aquifer, the water level rises up to the piezometric head. The wells in the confined aquifer are thus artesian wells operating under the piezometric pressure of the aquifer. If the piezometric surface at any well is above the ground surface, the water level in the well rises above the ground surface (Fig.7.5).

In such a situation, the well is known as an artesian flowing well. For example, the wells located in the Neyveli area in Tamilnadu are of this type. The formations around Neyveli are Cuddalore sandstones and clays of Miocene age. Miocene formations in India are known to operate under confined conditions with considerable piezometric head.

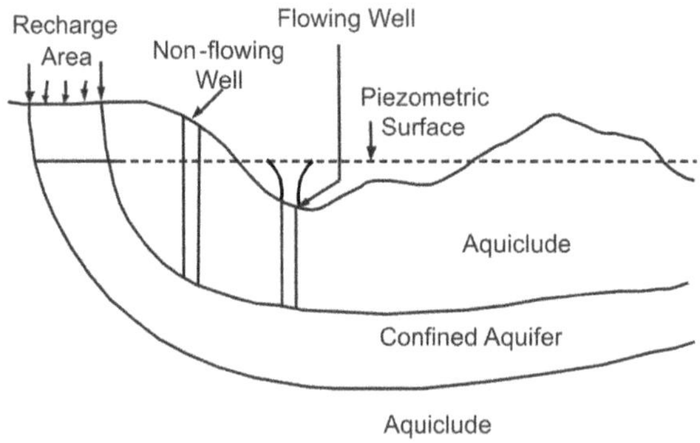

Fig. 7.5 Confined aquifer.

It is important to note that the water level in this case is not called water table but is known as the piezometric surface. Water is drained into the well under the piezometric head of the aquifer. The aquifer supplies water to a well by compaction of the aquifer. Unlike the unconfined aquifer, the confined aquifer stores less water and its storage is replenished at any place due to its piezometric head.

Leaky aquifer (Fig. 7.6)

If an aquifer is confined on either side by aquitard instead of aquiclude, then it is not truly confined in the sense that some flow of water takes place through the aquitard into the aquifer. If a well situated in such an aquifer is pumped, the contribution of water is not only from that aquifer but also from the neighbouring aquifers above or below with the flow taking place through the aquitard (Fig 7.6).

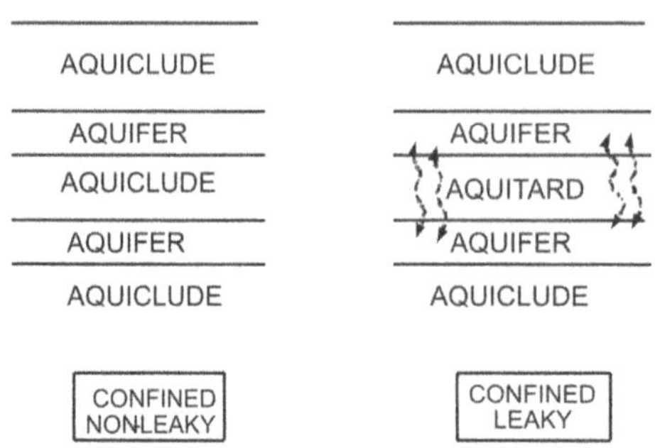

Fig. 7.6 Non-leaky and leaky confined aquifers.

Gangetic aquifers are typical examples of the leaky type. Sand and gravel horizons function as aquifers with the intermittent silty clay as the aquitard. The aquifers have hydraulic continuity with each other through the semipervious aquitard horizons.

Occurrence of Water in Various Lithological Formations

Igneous and Metamorphic Rocks

These two categories are similar as far as groundwater occurrence is concerned. The primary porosity is very low. It is as low as 8 to10% and sometimes even lower. Only in the basalts, the primary porosity is relatively higher due to the vesicles present. Joints, fractures and faults contribute to the secondary porosity. The permeability in a jointed rock varies with the extent of jointing in the rock. The joints decrease in their frequency with depth. Hence the permeability in a fractured igneous or metamorphic rock decreases with depth and ultimately the rock is hard and compact with practically no porosity or permeability. The top portion of the rock close to the surface is normally weathered. The fractured zone along with the weathered portion is the aquifer zone. However, as already indicated, the productive aquifer horizon is the portion of the aquifer below the water table.

Occurrence of impervious dykes and sills contributes to the groundwater localization. Further, the rock adjacent to the dyke or a sill has more fractures imparting added porosity and permeability.

The granites and gneisses in Peninsular India belong to the precambrian age. Average yield of wells in granites and gneisses of A.P. and Karnataka is reported to be around 350 l / min, where the rock is extensively fractured, increase in the yield up to about 1000 l/min has been the case. The permeability reported is of the order of 1.0 m/day. In Rajasthan however, the yields from the wells located in the weathered granite have been observed to be much less (around 50-60 l/min). Granites with joints rank as better aquifers compared to the gneisses. The schists are highly impervious and hence are rated as poor aquifers.

The basalts of the Deccan trap region are reported to be relatively good aquifers with yields ranging from 20 to 100 m^3/day. This is due to the primary porosity in the form of vesicles and also due to the occurrence of interbedded permeable sedimentary formations in between the individual lava flows. In addition, the rocks are jointed and also weathered in the upper horizons. The Deccan traps extensively occur in Maharashtra with their continuity into the adjoining states of A.P., M.P. and Karnataka. The permeability has been reported to be around 1.5 m/day.

Sedimentary Rocks

This group includes sandstone, siltstone, shale, claystone and limestone. The porosity in a sandstone varies with cementation. In sedimentary rocks horizontal permeability exists due to stratification. Just as in the case of the igneous and metamorphic rocks, joints and fractures in the faulted regions contribute to their better performance. Claystone, clay and shale have low permeability values and are poor aquifers. Coarse-grained sandstones and partly consolidated sandstones perform better as aquifers.

Limestones are better aquifers as in several cases, solution cavities are present within the same. Due to the selective dissolution, large opening are present in the limestone terrain underground. Development of the cavities is controlled by the composition (pure calcium carbonate and less of silica), extensive joints and absence of other interlayered formations like shale or clay. Often these openings extend to considerable distance in the subsurface. Mapping the geometry of these cavities is thus important during the exploration of the limestones. The limestones with the solution cavities are termed as karstic limestones.

High permeability values up to around 7.00 m/day are reported in the limestones. Occurrence of springs in a limestone is common in the areas where the limestone is confined within the impermeable horizons on either side and the flow takes place largely only within the limestone. Hence where such limestones intercept the ground surface along the slopes in particular, seepage of groundwater in the form of springs is a common feature.

Unconsolidated Sediments

Alluvial deposits, aeolian deposits and coastal sediments belong to this category. In general, these formations have high porosity and considerable permeability.

The alluvial deposits are the sediments deposited by a fluvial (water) agency. They include gravel, coarse sand, fine sand, silt and clay. Their deposition is controlled by the fluvial flow parameters like the velocity of flow, slope of the channel, intermixing with the load of the tributaries at their confluence with the master stream and finally at the deltaic regions at the confluence region with the sea.

The aquifers are composed of gravel and sand (sometimes with silt). They are interlayered with impervious clay horizons resulting in the confined nature of these aquifers. Where these confining layers are partly permeable (aquitards), these confined aquifers function as the leaky confined aquifers. This is the typical situation in the Gangetic basin. The Gangetic sediments are fluvial in nature. A gradation of size is conspicuous with the coarsest near the foothills of Himalayas grading into the finer sand and silt sizes in the Gangetic Plains. A typical cross section from North to South across the basin reveals interesting nature of the hydrogeology of the Gangetic terrain. The sediments coarsest in nature and exposed on the surface in the Northern reaches act as the recharge sources for infiltration of water and snow melt into the subsurface. This zone is termed as the Bhabar zone. These coarse deposits grade into sand and silt with the clay deposited during the intermittent flood seasons in the geological period. The slope suddenly increases in region immediately south of the foot hills. In this region the ground level being lower than the piezometric surface, the wells are of artesian flowing nature. Very high yields (free flows up to 250 m^3/hr), are reported in the wells within the region. This artesian zone is termed as the Terai zone. The boundary between the Bhabar and Terai zones is known as the spring line.

A relatively flat terrain designated as the Gangetic Plains follows the Terai zone to its south. This region has thick alluvial deposits composed of gravel, sand and silt alternated with clay layers. The gravel and sand horizons are the productive aquifers. The aquifers are either confined or leaky in nature. The wells operate as artesian (non-flowing) wells with considerable piezometric pressures. Since the deposits are unconsolidated and also the coarser layers are to be tapped for groundwater, the tube wells are gravel packed and the housing pipe having slots across the aquifer layers.

The water quality has been reported to be varying with depth in the Gangetic Plains. The dissolved solid content increases in the water drawn from deeper aquifers. This is due to the increase in geothermal temperature with depth and the slight increase in the temperature of the water with depth aids in the enhanced solubility of the dissolved constituents.

Coastal Deposits : These include deltaic sediments very close to the coast. As already indicated the Miocene deposits composed of sandstones on the east as well as the west coasts are good aquifers.

Aeolian Deposits : These are deposited by wind and composed of loose unconsolidated sand and loess. Sand deposits have both porosity and permeability while loess has porosity with extremely low permeability. Typically in the desert areas of Rajasthan, these deposits act as poor aquifers. Water quality is also poor.

Groundwater Exploration

Exploration for groundwater involves the mapping of potential aquifer horizons to be tapped for groundwater development. In hard consolidated rock formations, the jointed and fractured part is the aquifer horizon. Location of water table through field investigations helps to define the saturated zone of the aquifer since the openings in the rock below the water table are saturated with water (Fig.7.4). However in the unconsolidated formations like alluvium which contains alternate layers of pervious (sand and gravel) and impervious (clay) material, exploration for ground water involves finding out not only the nature of the individual layers but also their respective thickness values and depths of occurrence. Ground water quality assessment is also an important part of exploration operations

In the exploration for groundwater, several surface and subsurface investigations are carried out. Details regarding the methods and interpretation of data are indicated in chapter 6.

Geological mapping of the area forms the initial step. An integrated map is prepared based on the field data as also from the interpretation of aerial photographic and satellite data. Digital image processing of satellite data enables numerical manipulation of the digital data to obtain regional geological information and also relevant details for groundwater exploration. This involves application of various enhancement techniques to detection of features like lineaments zones and feasible areas for groundwater development. Multispectral data (of more than one band) is utilized in this process. Various steps adopted for groundwater exploration are indicated in Fig. 7.7.

In addition to the geological and structural features in the area of investigation, geomorphic classification is also carried out indicating various geomorphic features such as structural hills, residual hills, pediments, pediplains, valley fill and linear ridges. A final groundwater prospective zone map is prepared through integration of all the hydrogeological and hydrogeomorphic details.

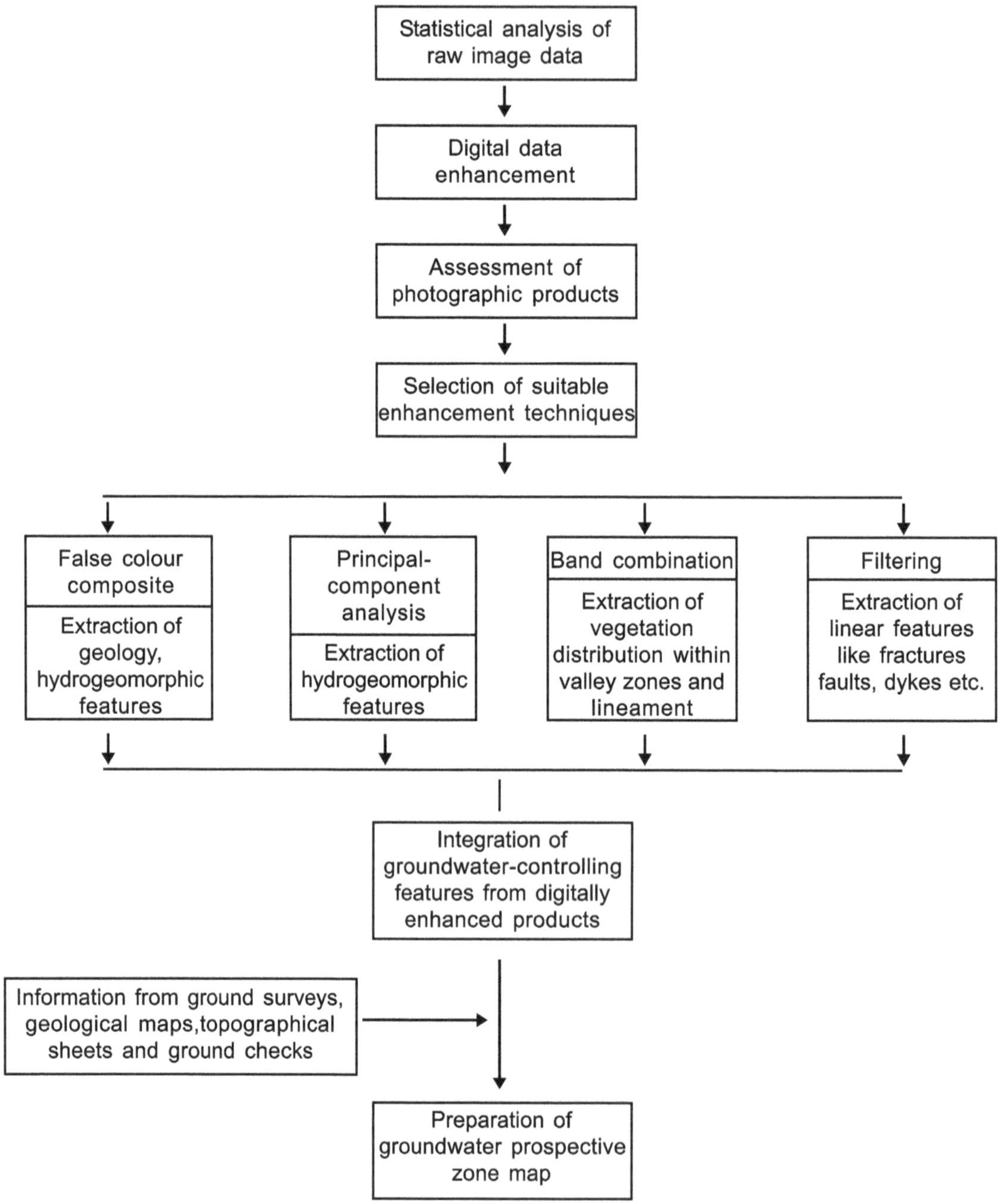

Fig. 7.7 Flow chart of digital enhancement operation involved in groundwater exploration. (Krishnamurthy, Manavalan and Saivasan, 1992)

As an example, typical groundwater prospective zone map prepared for an area with hard rocks is presented in Fig. 7.9. The satellite image (FCC) produced through enhancement using Landsat Thematic Mapper (TM) bands 3,5 and 7 is indicated in Fig.7.8. The area, located in Raichur District of Karnataka , is categorized as drought-prone. The area is composed of granites, gneisses, amphibolite and schists. The area is structurally deformed. Dolerite dykes occupy two sets of fractures. As can be seen in the figure, the limbs of a fold are displaced by a fault in the NE corner of the area.

Fig. 7.8 Map indicating geological and geomorphic details obtained from Landsat enhanced data in parts of Raichur District of Karnataka. (Krishnamurthy, Manavalan and Saivasan,1992)

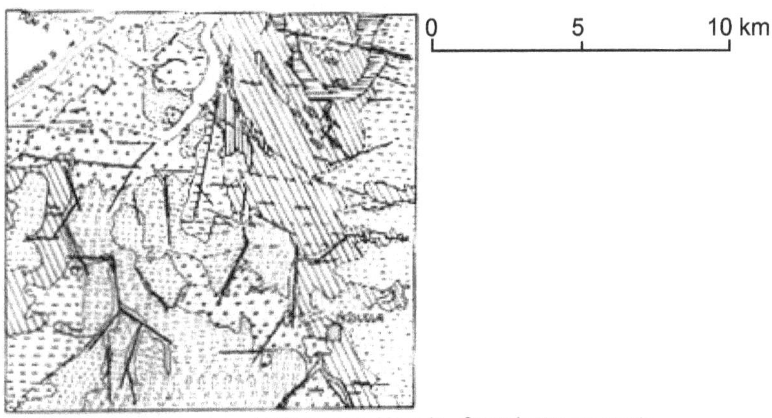

Hydrogeomorphic unit	Groundwater prospects	Hydrogeomorphic unit	Groundwater prospects
Pedplain over granite	Good	Resclal hill	Run off zone poor
Pedplain over onless	Good	Structural hill	Run off zone poor
Pedplain over SO_4	Moderate	Dyke	Acts as bearer CA ground water movement
Pedplain with black soil cover	Moderate quality	Moisture bearing lineament	Good to excellent
Pedment over granite	Poor	Dry lineament	Poor at sallow CCP
Pedment over gness	Poor	Fault line	
Valley hill	Good to excellent	Tank	

Fig. 7.9 Groundwater prospective zone map of the area in Raichur District (Krishnamurthy et al, 1992).

Remote sensing satellite data also enable mapping groundwater salinity. In this regard, remote sensing offers an excellent alternative to electrical methods of exploration. The surface expression of the soil and vegetation in a terrain can be analyzed distinctly in terms of salinity on a satellite image. A study on the basis of vegetation is valid since the scattered vegetation on the ground will appear more dense on satellite images, due to smaller scale (Bajpai and Gokhale, 1991). Favourable growth conditions for vegetation prevail in areas with shallow groundwater depth within about 10 m.

A methodology for the mapping of groundwater depth and salinity has been presented by Srivastava, Tripathi and Gokhale (1997). For this purpose an index known as the Green Vegetation Index (GVI) has been used which can be expressed as

$$GVI = [\{(NIR - R) / (NIR + R) \times 127\} + 128]$$

where NIR and R are the spectral reflectances of the near-Infrared and red bands respectively.

A reflectance zonal map for salinity in an area around Banda within the Gangetic Plains was prepared on the basis of GVI, using IRS -1B LISS-II data in bands 1 to 4 (Srivastava, Tripathi and Gokhale, 1997). The sequence of operations adopted in the analysis is indicated in Fig. 7.10. The reflectance zonal map with different salinity zones obtained on the basis of interpretation is presented in Fig. 7.11 The reflectance response in GVI image in response to soil, vegetation and hydogeological variations is indicated in Table 7.1.

Table 7.1 Reflectances in GVI image a function soil, vegetation and groundwater (Ahmad, 1984)

Zone	Reflectance zone	Image indication	Hydrogeological conditions
1.	Low salinity zone	Large bright patches representing dense vegetation.	Depth to ground water within 30 m. Normal salinity. Chloride content less than 180 ppm.
2.	Medium to low salinity zone	Scattered small bright patches with sparse vegetation	Depth to ground water less than 20 m. Chloride content between 200-180 ppm.
3.	High salinity zone	Dark patches due to absence of dense vegetation	Depth to ground water between 1.5 to 7.22 m. Chloride content more than 210 ppm.
4.	Ravenous / sandy zone	Dark patches along the surface streams	Depth to ground water very near to the surface.

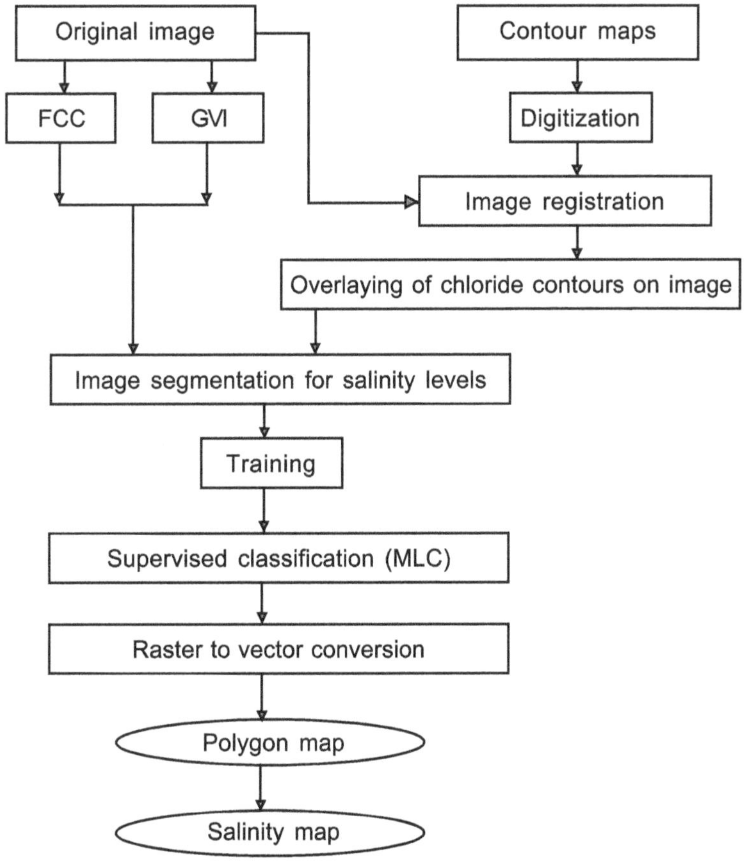

Fig. 7.10 Sequence of operations for delineation of salinity zones. (Srivastava, Tripathi and Gokhale, 1997)

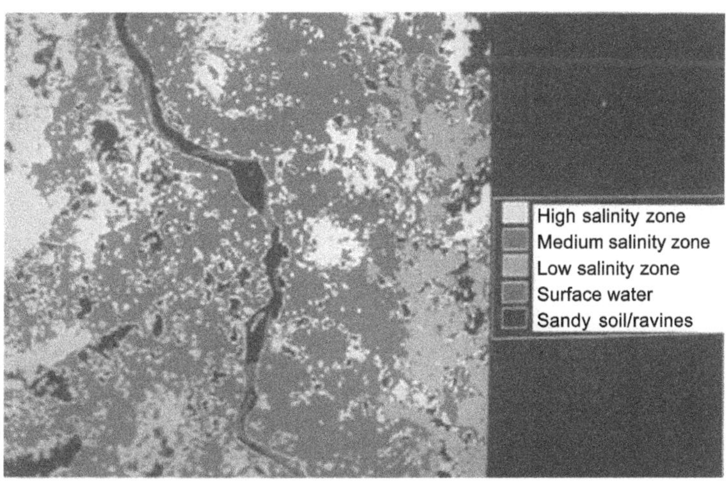

Fig. 7.11 Reflectance zonal map of Banda (U.P.) area indicating zones of differing groundwater salinity (Srivastava, Tripathi and Gokhale, 1997)

A regular variation has also been established between GVI on one hand and the water table depth as well as chloride concentration on the other. It has been reported by these authors that the salinity zones thus delineated are in close agreement with the reported chemical analysis of groundwater.

Geophysical Investigations

Geophysical exploration for groundwater involves :

1. Location of aquifer horizons
2. Water table delineation
3. Estimation of groundwater water quality

In hard-rock areas, the weathered and jointed portions of the formations function as aquifers. In such cases, as already explained, the aquifer portion below the water table is the productive aquifer horizon. Hence, not only the thickness of the aquifer is important but the thickness of the productive horizon is to be established during the investigations. In the case of confined aquifers, the thickness of the aquifer and the nature of the confining strata are important.

Among the geophysical methods, electrical resistivity and electrical well logging are the most widely used in groundwater exploration. Seismic refraction is also used for recognition of various formations, as well as their respective thickness values. In hard rocks, the seismic wave velocities in weathered and jointed portions are different compared to their saturated counterparts and also massive rocks. Details on the principles and applications of geophysical methods are presented in chapter 6.

Based on the subsurface data obtained from seismic and electrical methods together with the borehole logs, aquifer geometry is reconstructed. Although borehole data are at discrete points, the intermittent area is covered through geophysical methods and a continuous correlation of formations at different depths is carried out. Fence diagrams or panel diagrams are drawn. Typical fence diagram is presented in Fig 7.12.

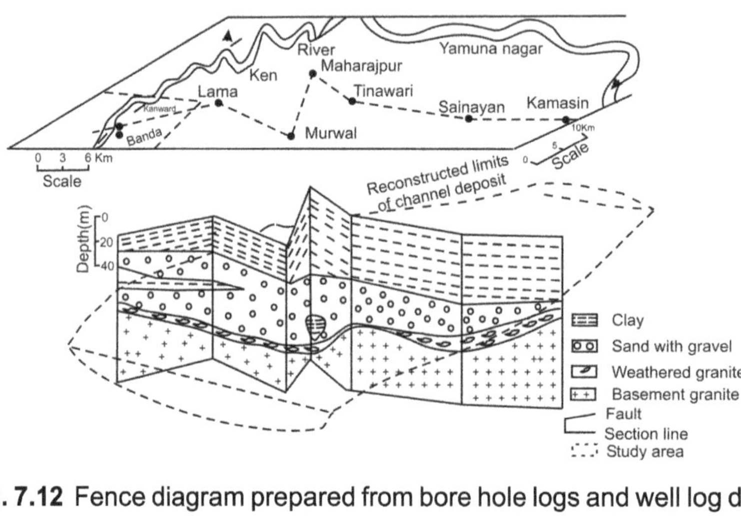

Fig. 7.12 Fence diagram prepared from bore hole logs and well log data. (Srivastava, Tripathi and Gokhale, 1999)

Fence diagrams are very useful in locating buried channels, which are potential sites for locating wells. Once the thickness variations of aquifer horizons are established, isopach maps (equal thickness contours) are drawn.

Electrical resistivity method provides adequate information in groundwater investigations. In addition to the nature, extent and thickness of different formations, an understanding of the groundwater quality can be achieved. Resistivity is the reciprocal of conductivity. So formations saturated with groundwater below the water table have more conductivity and hence lower resistivity values. With increasing salinity in the water, the resistivity further goes down in its value.

In addition to the induced potential measured in the resistivity method, the potential between the potential electrodes for different electrode spacing ('a') is also measured without any current input into the ground. This means the potential measured is the natural potential existing in the ground all the time and is not induced through current source. This method is known as the S.P. method. Natural potentials of considerable value exist in the ground with formations like clay, shale or any sulphide bearing ores.

Subsurface Electrical Techniques (Electrical Well Logging)

Electrode assembly is lowered in a drill hole and at regular intervals of depth and the readings are taken for material across on the wall of the well. Resistivity is computed if the field is an induced field. Just like the surface measurements as discussed earlier, the self-potential can also be measured without passing any current through the current electrodes. In Figs. 7.13 and 7.14, the details of the electrode arrangement being lowered in a borehole as also typical resistivity profile are indicated.

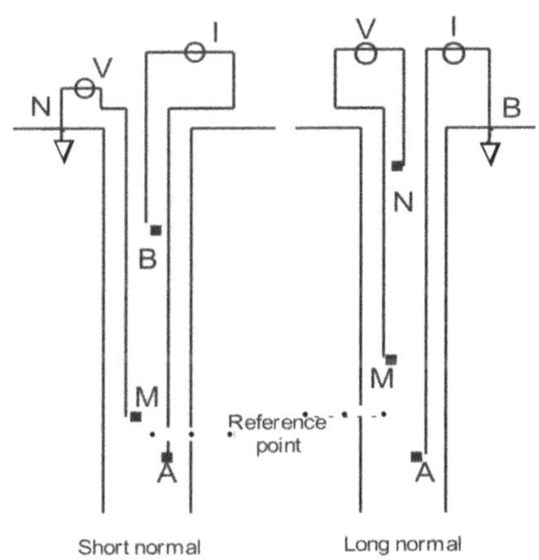

Fig. 7.13 Schematic diagram showing electrode spacings in resistivity logging of bore holes.

Electrical borehole logs provide very useful information. Not only are the aquifer horizons clearly indicated but also the lithological variation to depth is discernable on the record. For example the record in Fig. 7.14, obtained in alluvial sequence in the Gangetic plains, brings out clearly the grain size variations in the aquifer horizon, in addition to the demarcation of aquifers. Since alluvial aquifers are paleochannel deposits, a gradation of size from coarse at their bottom to finer sizes towards the top exists. As can be noticed, well logs provide continuous information right from the ground surface to the bottom of the hole. This is superior to driller's log, reconstructed from the samples obtained only at discrete depths and averaged to the depth intervals in between.

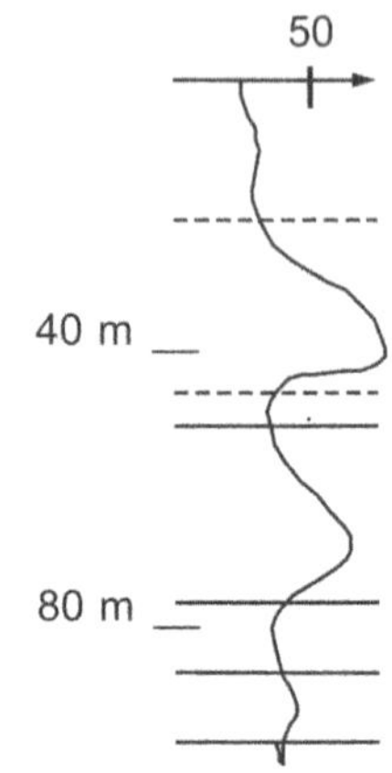

Fig. 7.14 Typical resistivity log.

Electrical logs are very useful for estimating the groundwater quality. In Fig. 7.15, a typical S.P. log is presented. The S.P value for the aquifer is noted from the log record. This is the departure of the curve from its base line across the horizon.

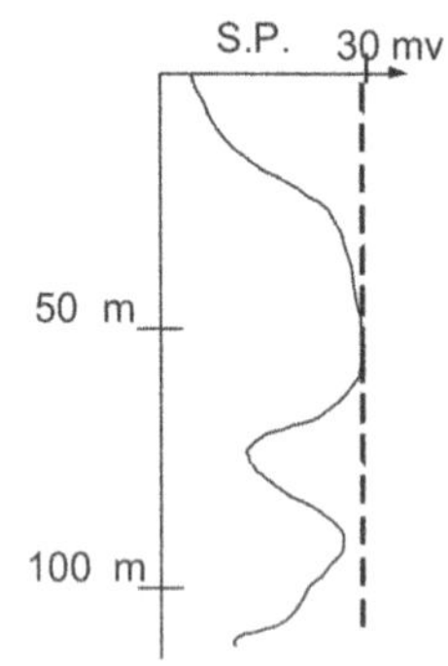

Fig. 7.15 S.P. log in a bore hole.

The salinity is calculated as follows :

$$\text{S.P. (in mv.)} = -k \cdot \log_{10.} (R_{mf} / R_w)$$

k = electrochemical constant

$= 60 + 0.133\ T\ (T \text{ in } {}^0F)$

R_{mf} = resistivity of the mud filtrate (measured in the field) (ohm.m)

R_w = resistivity of the formation water (ohm.m)

Resisitivity of the formation water is calculated by substituting rest of the values in the above expression. For resistivity of mud filtrate, the drilling mud is poured into a container fitted with electrode tapping facility on its side. Passing current through the two outer elechrode points, the potential is measured as in the surface measurements. Resistivity of the mud is then calculated.

Resistivity of the formation water is different from the resistivity measurements obtained in a surface resistivity survey. While the resistivity measured in a resistivity survey is the value for the formation along with its pore water, the resistivity of the formation water is due to the pore water only. Hence the resistivity values reflect the extent of groundwater salinity. The total dissolved content (TDS) of the formation water can be obtained from the resistivity-TDS calibration curve

Use of resistivity data for estimation of aquifer porosity

Porosity is related to the resistivity values (formation resistivity and formation water resistivity) through the following expression :

$$n^m = \rho_w / \rho$$

where n = porosity,

m = cementation factor (for sands it is around 2.0)

ρ_w= resistivity of formation water

ρ = in situ formation resistivity

Tracer Techniques in Groundwater Exploration

Groundwater movement in the subsurface is established using tracer techniques. The procedure involves the introduction of a tracer material in a well and monitoring the direction and velocity of the spread of the tracer in adjacent wells. To be a good tracer, it should not react with or pollute groundwater. It should be easily tracked even if it is present in low concentrations. Tracers are of different types :

1. Dyes like sodium flourescen , displaying colour in solution.
2. Salts like sodium chloride imparting conductivity.
3. Radioisotopes like tritium (H^3) with half-life of 12.4 years.

For proper monitoring, the observation bore holes or wells need to be closely spaced. The time of travel of the tracer between two wells of known spacing provides velocity. The velocity in turn can also be expressed as a function of hydraulic comductivity and porosity.

Use of Geographical Information System (GIS) in Groundwater Exploration

Geographical information system deals with management of databases. In ground water exploration, variety of data are collected on the area of investigation, based on which several thematic maps are prepared such as

- topographic map (Survey of India toposheets)
- geological map with all structural features (as described earlier)
- drainage map (with data extracted from toposheets / satellite imagery data)
- geomorphological map (based on geological map/satellite data)
- water table contour maps (based on water table depth measurements in the field)
- aquifer resistivity contour map (from surface resisitivity survey)
- aquifer thickness contour map(isopach map)(from resistivity/fence diagrams)
- rainfall contour map (isohyetal map) based on meterological data

Using GIS software, all the thematic maps are overlapped on one another to obtain an integrated picture of the area. On each of these thematic maps, different locations are ranked in terms of excellent, good, moderate or poor condition in terms of favorable groundwater situation using a ranking system (10 as excellent to 2 or 1 as poor). On integration, an integrated rank is worked out for the entire area. The sequence of operations is indicated in Fig. 7.16.

Finally a groundwater zoning map is prepared indicating different zones (excellent to poor) in the entire area. Delineation of potential zones for groundwater development (excellent to good zones) can thus be delineated for development of groundwater in the area. To validate the results, performance of the existing wells in the potential zones in the area can be checked.

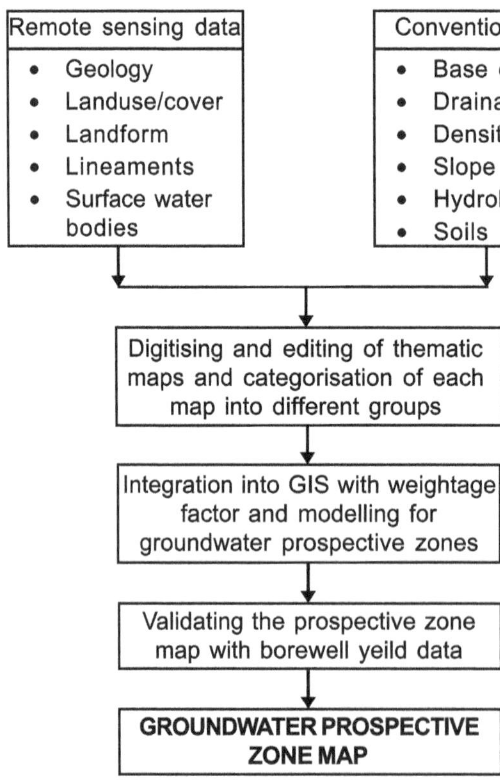

Fig. 7.16 Sequence of steps for the preparation of a groundwater prospective zone map using GIS

Groundwater in Coastal Areas

In areas close to a coast, fresh water availability is guided by the fresh water- salt-water interface (Fig. 7.17). The fresh water in storage below the water table floats above the salt water, depressing the interface to an equilibrium position. The density of fresh water is 1.00 gm/cc while that of saline water is taken to be 1.025 gm/cc. The fresh water-salt water depth relation, called the Ghyben–Herzberg relationship, can easily be worked out on the basis of density and thickness of fresh water lens up to the interface.

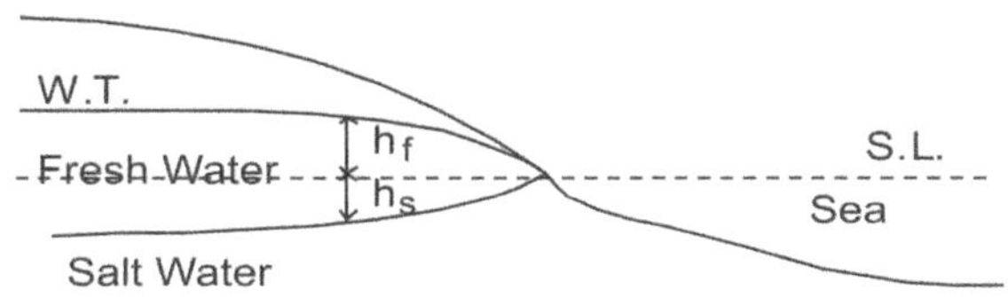

Fig. 7.17 Fresh water – salt-water interface in coastal areas.

The column of fresh water (Fig. 7.17) at point P is ($h_f + h_s$) and this suppresses the salt water from the sea level up to a depth of h_s. Thus based on the density relations, the salt water–fresh water equillibrium situation is expressed as

$$(h_f + h_s) \cdot \rho_f = h_s \cdot \rho_s$$

where ρ_f and ρ_s are densities of fresh water and salt water respectively.

Hence, $h_s = h_f \cdot \rho_f \,/\, (\rho_s - \rho_f)$

Since $(\rho_s - \rho_f)$ is 0.025, $h_s = 40\ h_f$

Around 40 units of fresh water accumulation depresses the salt-water interface by one unit of depth. Conversely, if 40 units of fresh water are removed, the interface comes up by one unit.

The depth of fresh water-salt water interface thus plays a significant role in the management of fresh water supply through wells in coastal areas. This interface can be easily located through electrical resistivity survey on the surface. Since the resistivity of the salt-water zone in a formation drastically falls down when the interface is intercepted during the survey, the contrast is very clearly discernable in the electrical resistivity profiles. Once the depth of the interface is obtained from the resistivity profile, care should be taken to limit the depth of a well in such a way that during pumping, the drawdown in the well should not bring up the salt-water interface into the well. Once the salt water diffuses into the well, it turns into brackish water well. The scheduling of pumping operations can be frequent and of shorter time durations, giving a scope for the recharge into the well in the intermittent time periods, resulting in the restoration of the interface back its original depth.

Aquifer Evaluation

The performance characteristics of the aquifers are evaluated on the basis of pumping tests on the existing wells (tube wells or bore wells).

Aquifers parameters used for characterizing aquifer productivity are :

- permeability
- transmissivity
- storativity
- specific yield of a well

Permeability has already been defined in the earlier portion of the chapter. While permeability (k) indicates the quantum of water passing through a unit area of cross-section of the aquifer under unit hydraulic gradient, transmissivity (T) is the term used for the flow of water through unit width across the entire thickness of the aquifer.

In confined aquifers since the entire thickness of the aquifer is drained under piezometric pressure, T = k.b (b being the thickness of the confined aquifer). This is not strictly valid in the case of the unconfined aquifer, which drains under gravity. The unit of 'T' is m^2/day ('k' is expressed in m/day).

Storativity (S) is defined as the volume of water the aquifer releases from or takes into storage per unit surface area of the aquifer per unit hydraulic gradient. It is a dimensionless parameter since

$$S = \text{Volume of water} / (\text{unit area}) / (\text{unit head}) = m^3 / (m^2.m)$$

Since the discharge from an unconfined aquifer is through decline of water table, the storativity is equal to the specific yield (specific yield is the ratio of volume of water that drains through the saturated portion by gravity to the total volume of the aquifer). In the case of a confined aquifer, since the supply of water is through compaction of the aquifer under piezometric pressure, it has low values of S (in the second or third decimal place). The smaller the value of S, the greater is the confining pressure in the aquifer, which denotes the increase in the artesian nature of the aquifer. Thus, value of S obtained in pump test data for an aquifer is indicative of its operating conditions, as unconfined or confined aquifer. On the other hand value of T indicates the productivity of the aquifer. Aquifers with higher values of T yield higher discharges.

Estimate of storage of groundwater in an aquifer can be arrived from storativity (S) as

$$V_w = S.\ A\ .\Delta\ h$$

V_w = vol. of water in storage

A = area of the aquifer

Δ h = average decline in the head

Pump Tests

Pump tests are conducted in the field to assess the productive nature of the aquifers. The aquifer performance is evaluated on the basis of the aquifer characteristics, namely permeability, transmissivity and storativity. In addition, the performance of a well after considerable period of its operation can be established taking recourse to pump tests. To precisely fix the reasons for a reduced yield from a well over a period of time, evaluation of aquifer loss and well loss through a step drawdown test is required. These losses enable an understanding of the contribution of the well and the aquifer in the reduction of the well yield.

When a well situated in an aquifer is pumped, the water table (in the case of an unconfined aquifer) or the piezometric surface (in the case of a confined aquifer) around the well is modified (Figs. 7.18, 7.19). The modification is in the form of a cone of depression deepest at the well and flattening away from the well spreading in all the directions. The shape of the cone of depression depends on the aquifer properties and the quantum of discharge from the well. The distance (R) on the ground surface from the centre of the well to the limit, up to which the cone of depression extends, is known as the radius of influence. For the optimal performance of wells in a well field, the radius of influence is helpful. The wells should be so located that their individual cones of depression should not overlap while the wells are being pumped. If they overlap, their mutual interference will reduce the effective yield of the respective wells. The lowering of the water level from the original static water level to the level on pumping is known as depression or drawdown. This is normally indicated in the groundwater terminology as "s".

Steady-state condition

A well is stated to be operating in a steady-state condition if the development of cone of depression over a period of pumping has reached an equilibrium state. Expressions for steady state operation of wells penetrating confined and unconfined aquifers have been developed by workers during early years for the estimation of aquifer permeability. The permeability thus obtained is known as the in situ permeability and truly represents the value for the aquifer as the test is conducted in situ.

1. ***Well penetrating a confined aquifer :***

The situation is illustrated in Fig. 7.18 for a well pumped for a discharge 'Q'. The drawdown, 's', is the difference $(h - h_w)$ between the level of water in the well on pumping and the static water level prior to the pumping. The discharge from the well is related to the aquifer permeability (k) as,

$$Q = k.\ i.\ A \text{ (Darcy's equation)}$$

I = hydraulic gradient and A = area of cross-section of flow into the well.

$$Q = k.\ i.\ (b.2\Pi r_w) = k.\ b.2\Pi.\ r_w\ .\ (dh / dr)$$

b = thickness of the aquifer and r_w = the radius of the well.

Integrating between the limits r_w (with h_w) and R (with h),

$$Q = \{k.2\Pi.b.\ (h - h_w)\} / \ln (R / R_w)$$

$$= k.\ 2\Pi.b.\ s / \ln (R/r_w)$$

Thus permeability can be estimated if the values of Q, R, r_w and s are known.

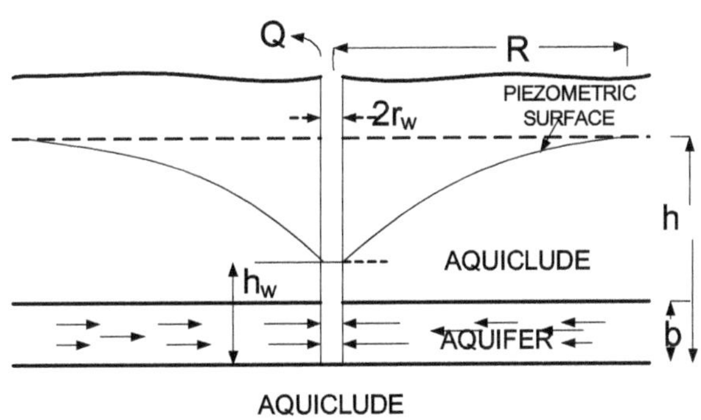

Fig. 7.18 Pump test on a well in a confined aquifer.

2. ***Well penetrating an unconfined aquifer :***

The situation is illustrated in Fig. 7.19 The basic difference between this case and that of the confined aquifer is that in the present situation, the cross-sectional area of flow between the limits R and r_w varies in the saturated aquifer horizon.

Thus the drawdown – discharge relation can be worked out through integrating the basic Darcy's equation, as before, but with the limits as applicable.

The final equation can be written as

$$Q = k.\Pi. (h^2 - h_w^2) / \ln (R/r_w)$$

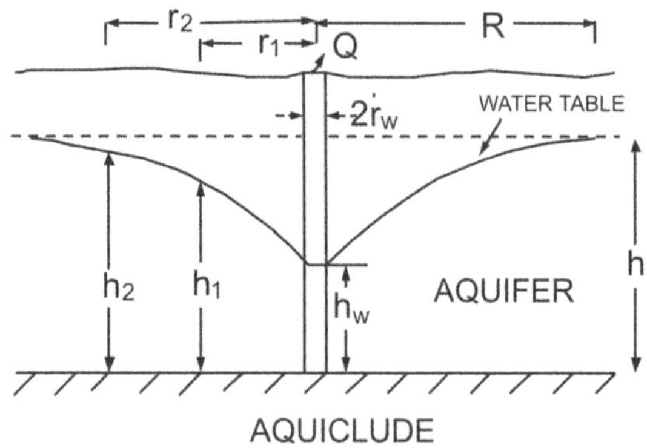

Fig. 7.19 Pump test in an unconfined aquifer.

Steady-state conditions are attained after several hours of pumping and thus in practice, the tests conducted normally do not pertain to this situation.

Non-steady State Condition

For a pump test conducted with a well pumped for a limited period of time, the steady-state equation cannot be applied strictly as the well is working under non-equilibrium conditions. Theis (1935) has developed a drawdown-discharge relation incorporating a well function. Theis equation applies to a confined aquifer condition.

The drawdown- discharge equation is

$$s = (Q / 4\Pi T) . W(u)$$

where s = drawdown

T = transmissivity (= k.b)

$W(u)$ = well function

$u = (r^2 S) / 4T.t$

r = distance from the main well to the observation well

t = pumping period

Theis has reported a type curve prepared with different values of W(u) as plotted against the corresponding values of u. This curve is known as the Theis non-equilibrium type curve for confined aquifer analysis.

Time-drawdown data as obtained in the field are plotted on a log-log plot of the same log-cycle as the type curve. The aquifer parameters T and S are obtained by curve-matching of the field curve with the type curve.

Details of the procedure for aquifer parameter estimation :

Following are the steps involved for pump tests conducted under non-equillibrium conditions of pumping (for shorter duration of pumping during which time steady state is not attained).

1. The static water level before pumping the aquifer is noted. (this corresponds to the depth of water table in an unconfined aquifer and the piezometric surface in a confined aquifer).
2. The well is pumped through at a particular controlled rate.
3. Water-level measurements are made (using an airline / wetted tape / electrical measuring line) at different intervals of time. In the initial stages, since the lowering of the water levels in the well is at a faster rate, readings are to be taken at closer time intervals. These intervals can be of larger duration if the decline takes place at a lower pace. The difference between this water level and the static level measured just before the commencement of the test is known as the drawdown or depression in the well.
4. The yield (Q) from the well is measured using a V-notch / L-scale/ orifice bucket or any other means.

   The results are tabulated as follows :

   Static water level in the well = ... (m). Distance (r) / of observation well = ... (m).

Time (min.)	Water level (m)	Drawdown (m)
________	________	________
________	________	________
________	________	________

Plotting and interpretation of field data :

1. The drawdown values (s) are plotted (on a log-log plot) versus time (t). The modulus of the graph paper used should be same as the type curve available.The plotting of field curve should be on a transparent paper to facilitate matching with the type curve.
2. The field curve is matched with the type curve, taking care to see that the corresponding axes (the x- and y- of both graphs) are kept parallel (Fig. 7.20).

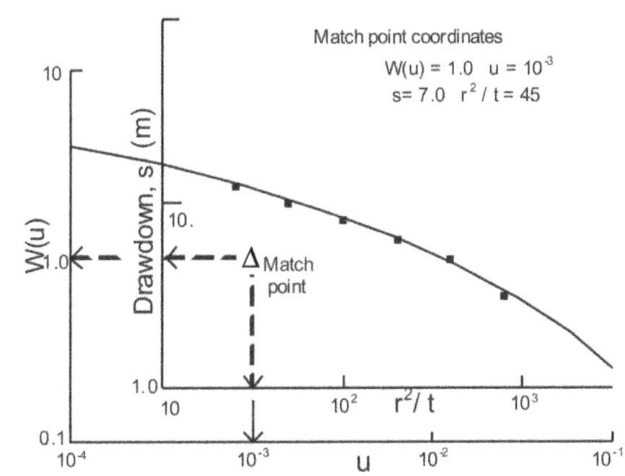

Fig. 7.20 Matching of field curve with Type Curve (non-equilibrium equation).

3. The Theis type curve for confined aquifer is used. At the perfect matching position, all the four coordinates (the x and y for the field curve, the x and y for the type curve) are noted, for a chosen match point on the curves.

Coordinates of the match point :

For type curve :	W(u)	along y-axis	,	u	along the x-axis
For field plot :	s	,, ,,	,	r^2/t	,, ,,

Using the Theis equation, the aquifer parameters are estimated as follows :

$$T = Q \,.\, W(u) / 4\,\Pi.\, s$$

$$S = 4T.t.u / r^2$$

where ' r' is the distance of the observation well in which drawdown measurements are made (r will be the radius of the pumped well if observations are made on the same well instead of an observation well).

The S values will be small (in the second or third decimal place) in the case of a confined aquifer.

If a Theis type curve is not available, T and S can be estimated using the Cooper-Jacob approximation of Theis equation. The plotting and interpretation for the same are :

1. Plot the drawdown (s) and time (t) data on a semi-log paper (Fig. 7.21).

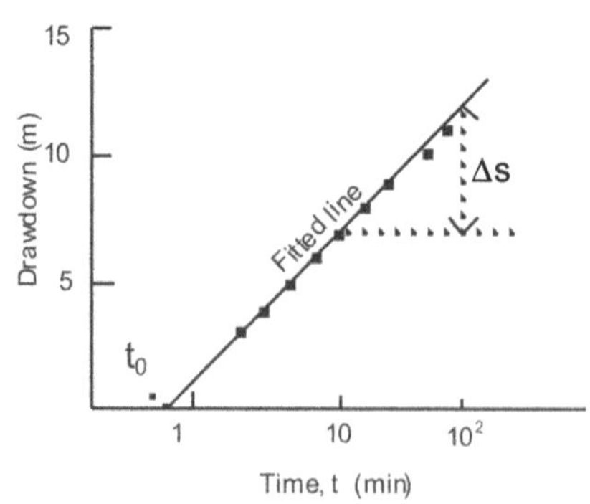

Fig. 7.21 Cooper-Jacob straight-line method.

2. Note the drawdown (Δ s) for one log cycle of t (x-axis)
3. Calculate the aquifer parameters 'T' and 'S' as follows :

$$T = 2.3\, Q / (4\,\Pi.\, \Delta s) \qquad S = 2.25\, T\, t_0 / r^2$$

Q = pumping rate, t_0 = time where the straight line plot intersects the x-axis.

r = distance of measuring point.

Evaluation of a Leaky Aquifer

As already explained, the contribution (Q) from a leaky aquifer is both from its storage as well as the contribution into the aquifer from an adjacent aquifer through an aquitard (Fig 7.22). The pump test data collection is similar to the procedure outlined above for the unsteady state (non equilibrium conditions). The field curve plotted with 's' (drawdown) versus 't' (time), cannot be matched with a Theis type curve fully as in the earlier case with a confiner aquifer test data, since the drawdown in the aquifer does not truly reflect the yield from the well. The field drawdown curve in the present case departs from the type curve for its latter half of the data plots. The field curve is relatively flat compared to the Theis type curve, essentially due to the leakage. In fact, this is the indication of a leaky aquifer.

Curve matching is then to be done using the Walton's leaky aquifer type curves and the field curve will certainly match with one of the curves on the type curve set (Fig. 7.22). As earlier the four coordinates are noted for the chosen match Δ point on the curves in the matched position.

The four coordinates are : W (u, r/b), 1/u on the matched type curve axes s and t on the axes of the field curve.

In addition, the value of r/b of the type curve that was matched is also to be noted, which provides information on the vertical permeability value of the aquitard.

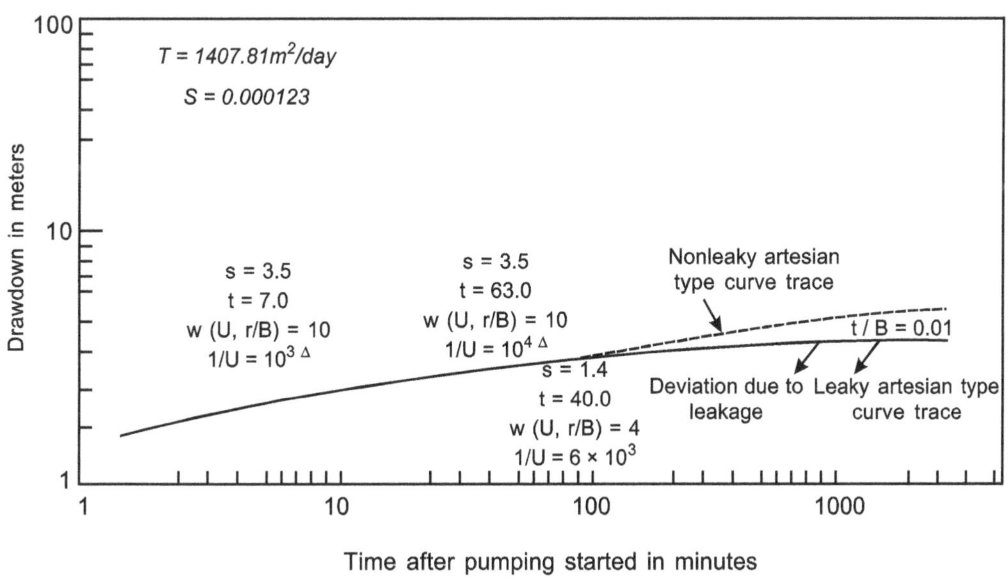

Fig. 7.22 Estimation of aquifer parameters for a leaky aquifer.

Using all the above, the estimation of aquifer parameters is as follows :

The four coordinates (two on the type curve and two on the field curve) for the match point in the matched position are noted. They are W(u, r / B), u, s and t. Using these values, the aquifer parameters T and S are estimated as follows :

$$s = Q/4\Pi T \{W (u, r / B\}$$

where W (u, r/B) is the well function
s is the drawdown
T is the transmissivity
Q is the yield from the test well

The value of T is obtained from the above expression by substituting the other parameter values as obtained from curve matching.

Similarly the storativity (S) can be estimated from the expression :

$$u = r^2S / 4Tt$$

where r is the distance of the observation well (from which the drawdown data is collected) to the main well. The values of u and t are the ones obtained as match point coordinates. The value of T has been estimated in the previous step.

Each of the Walton's type curves for leaky aquifers corresponds to a particular r/B value, which is indicated on the type curves. Once the field curve matches with a type curve, its value of r/B also has to be noted.

$$r/B = 1/ \sqrt{[T / (K'/B')]}$$

Since r/B and T are known, K'/B' can be estimated. K' is the vertical hydraulic conductivity of the aquitard (the confining formation) and B' is its thickness. If the B' is known (from the drill log of the well), K' can be estimated. Thus the leaky status of the aquitard can be determined.

Springs

A spring is a continuous flow or seepage of water on the ground surface. Springs are located at the points of intersection of groundwater table with the surface of the ground.

All the springs in nature are grouped into two main classes by Bryan:

1. *Gravitational :* water flow resulting from gravitational condition
2. *Non-gravitational :* with flow under non-gravitational condition

Gravitational springs can be due to

- topographic depressions where water table intersects the surface (Fig. 7.23 A).
- openings in a confining bed above an aquifer (Fig. 7.23 B) : artesian springs.
- escape of water from solution cavities onto the surface (Fig. 7.23 C).
- contact of permeable with impervious formation (Fig. 7.23 D) : contact spring.

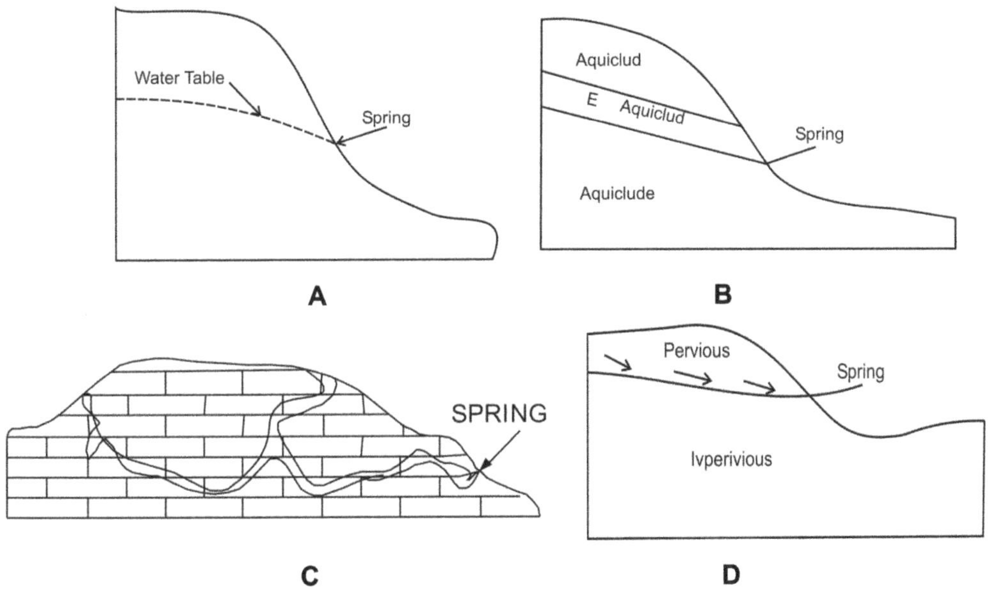

Fig. 7.23 Gravitational springs.

Non-gravitational springs (Figs. 7.24, 7.25) are mainly due to

- fractures extending to great depths with escape of hot water(thermal springs)
- volcanic vents and escape of volatiles.

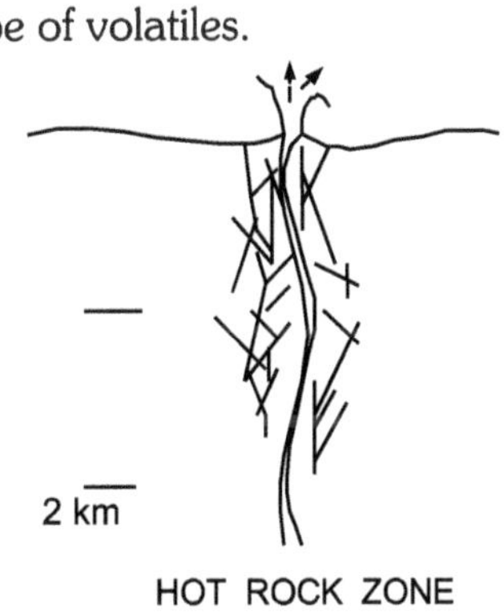

Fig. 7.24 Cross-section of a hot spring zone.

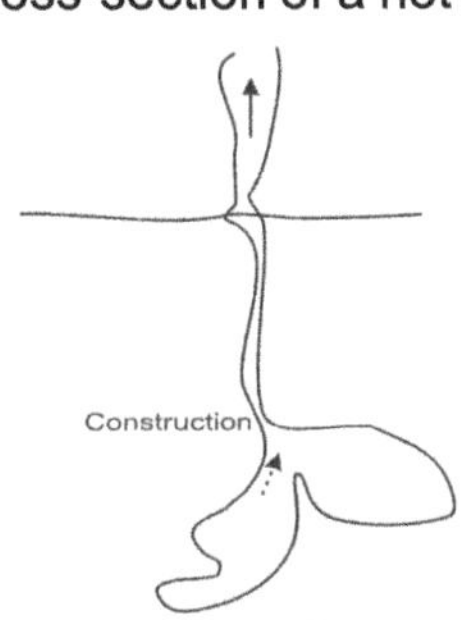

Fig. 7.25 Cross-section of a geyser conduit.

Thermal springs are potential sources of power generation. In India, around 400 thermal springs exist distributed within 7 major geothermal provinces (Fig. 7.26). Another major geothermal province has emerged in the Barren Islands (located at 116km ENE of Port Blair, Andaman Islands) after the recent volcanic eruption (1991). In the Western Himalayan region itself (Fig.7.27), there are about 100 thermal springs distributed around Puga (Ladakh district of J & K state), and Manikaran (Kulu, H.P.). The temperature varies from 47-98^{0}C. Estimated power generating capacity from all major Indian thermal springs is about 10,600 MW (Chandrasekharam, 2000).

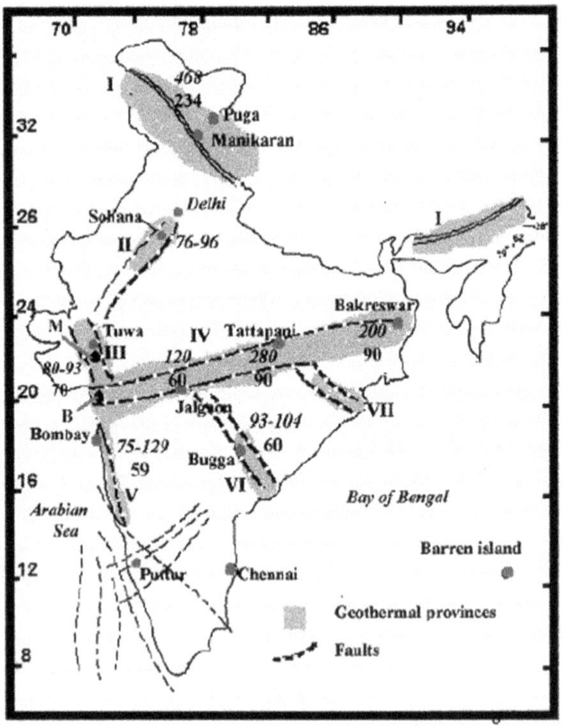

Fig. 7.26 Major geothermal provinces of India I : Himalayas II: Sohana III : Cambay IV : Sonata V : West Coast VI : Godavari VII : Mahanadi (heat flow values in italics, mW/m^2, geothermal gradients, ^{O}C/km) (Chandrasekharam, 2000).

Water from the thermal springs has varying high temperatures depending upon the path and the rate of flow from the subsurface. In a thermal spring the water temperature is at least 5 to 6 ^{0}C more than the atmospheric temperature. The geothermal gradient (rate of increase of temperature with depth) is around 1^0 / 50 m. Surface temperatures of the spring waters vary between 57^0 to 98^0 C. The reservoir temperatures have been estimated to be around 300^{0}C.The flow rate of water varies between 200 lpm and 1000 lpm. The temperature of water increases with deepening of fractures due to deformation resulting from an earthquake. For example, in recent years in Badrinath, an increase of temperature of the spring water has been observed after the UttarKashi earthquake of 1991.

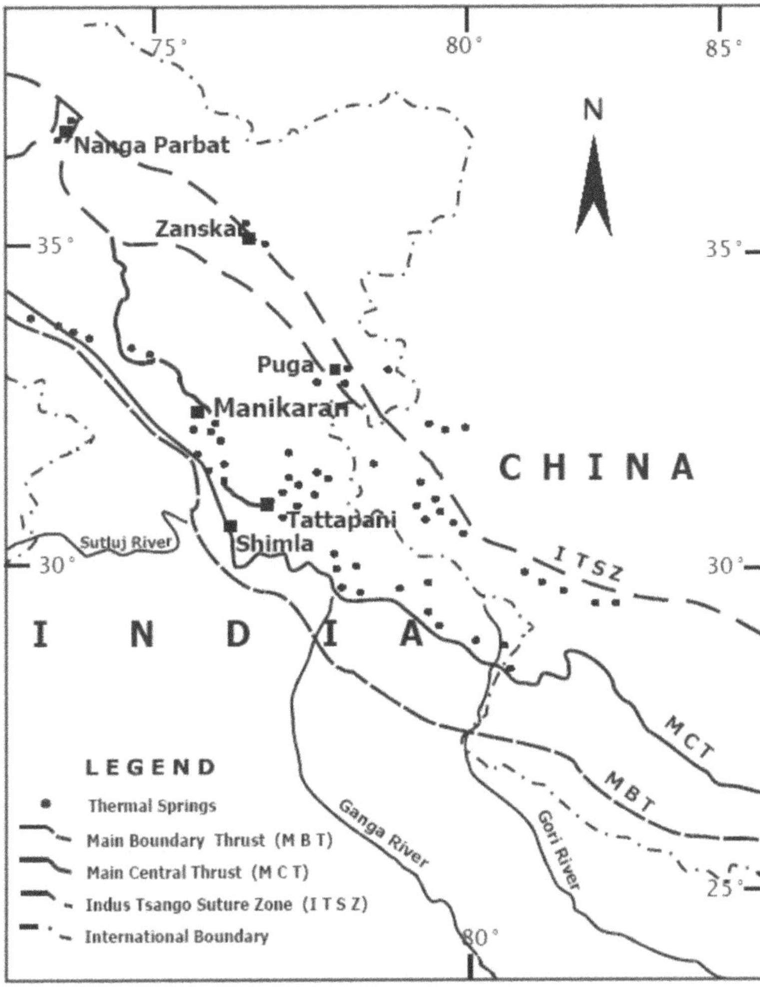

Fig. 7.27 Geothermal manifestations in N.W. Himalayas.
(Chandrasekharam, Ayaz Alam and Minissale, 2003)

If a hot spring erupts periodically issuing out hot water and steam, it is known as a geyser. In the case of a geyser, the channel for the passage of superheated water with steam is constricted at places preventing free circulation of water to the surface as it happens in a thermal spring. Because of the prevailing high temperature at depth, the accumulated water boils. The resulting steam builds up pressure and forces up the accumulated hot water through the vent. Once this happens and the pressure is released, again the process repeats with the accumulation of water at the constriction. Steam and boiling water are issued at perfect time intervals from a geyser. Geysers are common in Iceland, Newzealand, Indonesia and the U.S.A. The Old Faithful geyser in the Yellow Stone National Park in U.S.A is a very popular one (Fig.7.28). Around 14,000 to 30,000 liters of boiling water with steam is released periodically at an average interval of about 75 minutes, reaching up to a height of 40 to 50 m.

Fig. 7.28 Old Faithful geyser in Yellow Stone National Park, U.S.A (Photo: Nagendra Kolluru).

Hot spring water is rich in dissolved chemicals of the rocks and this mineralizing water, rich in sulphur has medicinal effects. Volcanic springs with the discharge of various gases like CO_2, H_2S and SO_2 are called fumaroles.

On the basis of their discharge, springs were classified by Meinzer into different magnitudes :

Magnitude	Mean discharge
First	$> 10\ m^3/s$
Second	1 – 10 "
Third	0.1 – 1.0 "
Fourth	10 – 100 l/s
Fifth	1 - 10 "
Sixth	0.1 - 1.0 "
Seventh	10 - 100 ml/s
Eighth	< 10 "

Groundwater Provinces of India

A groundwater province encompasses regions with similarities of occurrence of ground water. As long as the similarity exists, these regions need not be adjacent to each other. Also the rock types in a groundwater province can be of different types with varying composition. Steps adopted for exploration, development and management of ground water are also more or less similar within the same province.

Eight groundwater provinces have been reported by Taylor (1959)

1. Precambrian crystalline province (Precambrian to Archaen)
2. Precambrian sedimentary province (Precambrian)
3. Gondwana sedimentary province (Late Carboniferous to Early Cretaceous)
4. Deccan Trap province (Upper Cretaceous to Lower Eocene)
5. Cenozoic sedimentary province (Eocene to Pliocene)
6. Cenozoic fault basins (Pleistcene to Recent)
7. Ganges–Brahmaputra alluvial province (Pleistocene – Recent)
8. Himalayan highland province (Archaen to Recent)

Hydrogeological details of the groundwater provinces :

1. Precambrian crystalline province

- Igneous (plutonic) and metamorphic rocks with surface weathering (up to around 20 m of thickness).
- The fractured and weathered zone is the aquifer horizon
- Due to restricted permeability, groundwater development is mostly through large-diameter wells with enough well storage, for irrigation purposes
- Water supply is also through bore wells
- Well discharges are low (around 25 m^3 / day)
- High flouride concentration in the granites of S. India

2. Precambrian sedimentary province

- comprises 4 structural basins (Cuddapah basin in A.P. , Raipur basin in Chattisgarh, Vindhyan basin in U.P., Bihar and M.P. , W.Rajasthan basin)
- Rock types are sandstones, shales and limestones
- Groundwater occurs in weathered zones and cavities in the limestones.
- The cavernous limestones yield more than 50 m^3/hr
- Sandstones and shales yield around 25 m^3/hr and 8 m^3/hr respectively.

3. Gondwana sedimentary province

- Includes the coal basins with sandstone and shale
- Distributed along Godavari, Mahanadi , Damodar , Narmada , Son valleys and parts of Rajasthan and Gujarat
- The sandstones are good aquifers
- Development is through dug wells (350 m^3/day) and tube wells (600 m^3/day)

4. Deccan Trap provinces

- Volcanic rocks (basalt) as layers of lava flows interbedded with sedimentary rocks
- Distributed extensively in Maharashtra with extension into A.P., M.P., Karnataka, Gujarat and Rajasthan
- Below these rocks, potential sandstone aquifers exist
- The rocks are weathered and jointed
- Yield from dug wells around 50 m^3 – 100 m^3/day

5. Cenozoic sedimentary province

- Sandstone and clay deposits along East Coast, the Warakali formations (Kerala) on W. coast and sedimentary formations in Cambay Basin (Gujarat)
- Also in the Krishna-Godavari interstream region and Cauvery-Paleru inter stream region
- Artesian aquifers with enough piezometric pressures – at places flowing wells like in Cuddalore sandstone around Neyveli
- Tube wells with large yields (around 150 to 200 m^3/hr)
- In Cambay basin, groundwater at depth is brackish

6. Cenozoic fault basins

- Three fault basins (Narmada , Purna and Tapti valleys) included
- Valley deposits comprise gravel, sand, silt and clay
- Thickness varies from 50 m to 150 m
- Sand and gravel are the aquifers with yields upto around 100 m^3/hr
- Groundwater in Purna valley is saline and not fit for use

7. Ganges–Brahmaputra alluvial province

- Very extensive covering several states from Punjab to W.Bengal
- Alluvium is composed of sand, silt and clay with gravel horizons.
- Thickness of alluvium is extensive
- Three distinct subdivisions: Bhabar, Terai and Gangetic alluvial plains from N. to S.
- Bhabar zone at the foothills of the hills serves as the recharge source.
- Aquifers in Terai and Gangetic Plains are artesian with considerable piezometric pressures.
- Wells in Terai zone are often flowing artesian wells

8. Himalayan highland province

- Extensive distribution all along the Himalayan ranges from Kashmir to N.E. India.
- Highly deformed (folded and faulted) rocks: Metamorphosed sedimentary formations, limestones and also granites in the upper reaches
- Occurrence of thermal springs within the Himalayan geothermal province. Around 100 thermal springs in the N.W. Himalayan region of Ladakh (J&K) and Manikaran (H.P.) and adjacent areas.

Problems of Groundwater in Civil Engineering practice

While groundwater is the most sought after resource for water supply, it poses problems to a civil engineer or an engineering geologist during the execution of construction work. Problems commonly encountered are as follows :

1. In tunneling operations, seepage (and sometimes sudden gush of water) into the tunnel opening. Proper lining techniques are called for in such cases. Groundwater problems in tunneling are detailed in chapter 9.

2. In excavations extending below the groundwater table or below the piezometric surface, dewatering has to be undertaken during the execution of the work. This can be achieved through a battery of wells appropriately spaced or through electro-osmosis.

   Classic example in India is the dewatering in the artesian aquifer system in Neyveli lignite area with very high piezometric head, much above the ground surface.

3. Seepage from the reservoir to the dam foundations endangers the dam stability, through an increase in the uplift pressure as well as through the slippage produced due to the lubricating effect of water. Foundation grouting is commonly done to stop the seepage in the dam site and also into the embankments from the reservoir. Details are presented in chapter 8.

4. Pollution of groundwater because of industrial effluents reaching the ground water level due to improper waste disposal.The details are presented in chapter 11.

5. Pore water pressures and slippage induced due to water seepage resulting in landslides. As explained in chapter 10, water is one of the main contributors to slope instability leading to landslides. In areas susceptible to landslides, release of pore water pressures can be achieved by providing adequate and effective subsurface drains to expel the water in the geological materials.

8

Engineering Geology in Dam Construction

Geological considerations play significant role for the location of dams and constituent reservoirs for impounding water. The type of a dam to be constructed is decided on the river valley characteristics and the geology. The foundation requirements for any dam are dependent on the site geology and the type of the dam.

Dams are constructed for impounding water. Various purposes of dam construction are:

1. To generate hydroelectric power
2. For providing irrigation
3. For flood control
4. For regulation and control of stream load
5. For water supply

Dams intended for more than one of the above objectives are termed as multi-purpose dams. In India, more than 90% of the dams operating are primarily for irrigation.

Dams are of different types. Either they can be totally of reinforced concrete or totally of earth materials or a combination of both.

Types of Dams

Gravity Dam (Fig. 8.1)

It is a heavy concrete structure. The weight of the dam, acting vertically, plays an important part in its stability. It holds water, impounded in the reservoir, by its sheer weight. The stability of a concrete dam depends on the pressure distribution. The different pressures acting are indicated in Fig. 8.1.

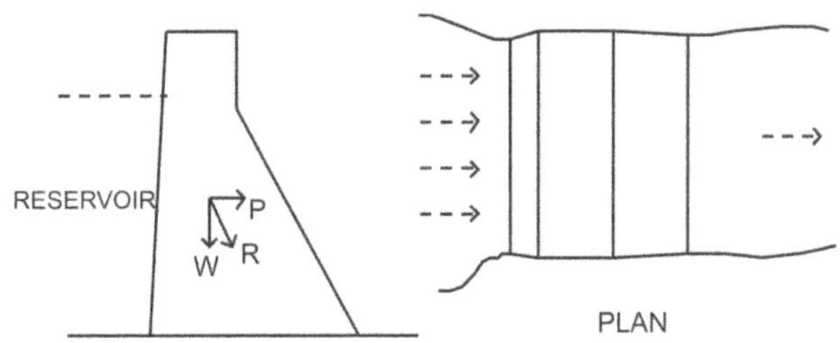

Fig. 8.1 Gravity dam.

For the dam to be stable, the resultant "R" (Fig. 8.1) of the vertical force, "W" (weight of the dam) and the lateral thrust, "P" (of the reservoir water) must be within the middle third of the dam. Besides, the formations below the dam in the dam site have to be very competent. Foundation treatment like grouting is adopted in case of any incompetent foundation material present.

Arch Dam (Fig. 8.2).

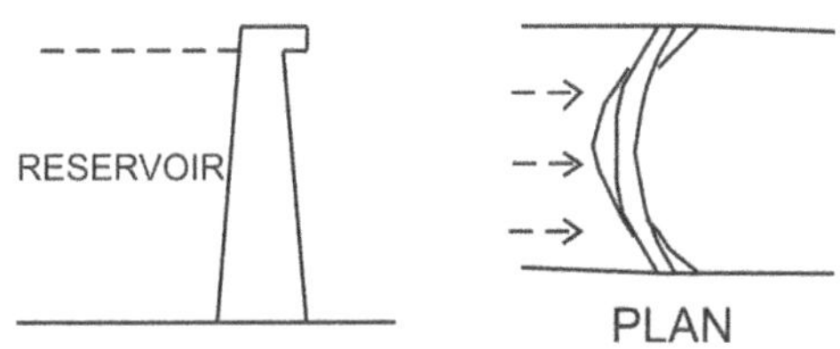

Fig. 8.2 Arch dam.

This type is preferred for narrow river gorges. The arch is convex to the upstream side. It can be single arch or multiple arches. The dam structure transmits force than resisting the same. It holds water due to its design features. Since a substantial part of the load is transmitted to the abutments, the formations constituting the abutments must be very competent. Arch dams need better monitoring.

Buttress Dam (Fig. 8.3).

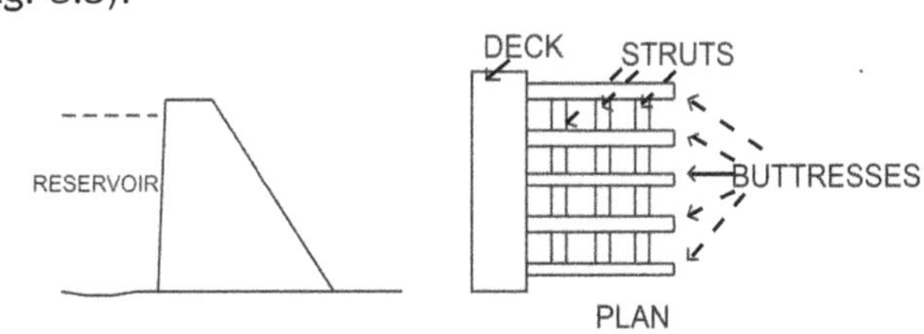

Fig. 8.3 Buttress dam.

As indicated in the figure, it has a concrete deck sloping upstream, supported by buttresses (or walls) placed at right angle to the dam axis. These buttresses are further strengthened by cross-walls known as struts. The entire system with all these components distributes the load over the foundation. Since the load distribution is over a wide area, even formations relatively weak are considered if other alternate better sites are not available.

Earth Dam (Fig. 8.4).

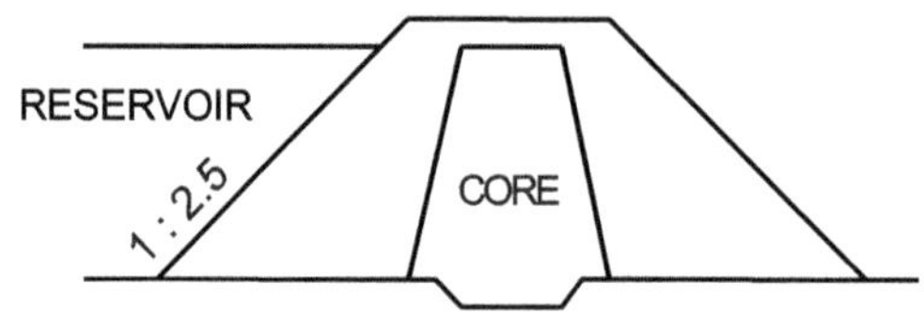

Fig. 8.4 Earth dam.

Constructed with earth materials, these structures are large in size and trapezoidal in shape. These are preferred in broad valleys and where the foundation material is weak. Because of the broad nature of the dam, the weight of the dam is distributed over a wide base and the force per unit area is consequently less. This is one of the main advantages in its choice. Another advantage is its adjustment to minor dislocations without failure. The earth materials used in its construction are gravel, sand, silt and clay. Rock-fill is also used. An earth dam with considerable rock-fill is termed as a rock-fill dam.

The side slopes are maintained at 1 in 2.5 or 1 in 3.0 and an impervious clay core (Fig. 8.4) is provided to arrest seepage across the structure. In addition to the core, filters are also provided to facilitate drainage and to release the pore pressures. On the upstream side, the slopes of the dam are paved with rock rubble and provided with an impervious blanket while on the downstream side, the slope is protected from erosion.

Location of a Dam

The ideal site for location of a dam should satisfy the following requirements:

1. A narrow river valley along with steeper side slopes
2. Stable slopes both at the dam location and along the reservoir sides, all the way up to the top of the slopes
3. Absence of weathered formations
4. Competent geological formations devoid of weak zones
5. Absence of clay and fractured material.
6. Absence of fault zones
7. Stability of formations below the dam and reservoir area after the water is impounded in the reservoir
8. Easy access and supply of materials for the construction of the structure

It may be mentioned here that all these attributes never exist in any one particular site. Consequently, appropriate foundation treatments are resorted to for making the dam site and reservoir area most suitable.

Geological Considerations

1. *Topography and geomorphology of the site*

As already explained, a narrow river valley with steep stable slopes forms the primary consideration. The channel deposit or the overburden preferably has to be of limited thickness with the occurrence of bedrock at shallow depths. Quite often, the valleys have talus and flood deposits along the slopes, thus giving a narrow appearance to the valley for an otherwise relatively wider river valley. During investigations, the valley status and soundness of its slopes has to be ascertained.

2. *Impact of geological structures*

The formations at the dam site should be dipping towards upstream or horizontal. This will counter the seepage, compared to the situation where the formations dip in the downstream direction. In addition, the foundations will have greater stability as the load is normal to the horizontal formations or formations with low dips. Faults or shear zones present in the formation result in the presence of weak zones with crushed material. Extensive jointing with intersecting sets in the rocks also threatens the safety of the structure with not only aiding seepage but rendering the formations incompetent. Presence of anticlinal or synclinal structures in the foundation material also contributes to the seepage problem. In the crest regions of anticlines, tensional cracks develop which pose threat to the competency of the deformed rock.

3. *Lithology of the formations*

Among the igneous and metamorphic rocks, granites, quartzites, gneisses and basalts are most suitable. Sandstones have primary porosity (intergranular porosity) and also are often interlayered with shales in a sedimentary sequence. While pure sandstones have good compressive strength, presence of interlayers results in the decreased strength. In the Siwalik sequence, the sandstone members are found to be artesian in nature. For example, in the Beas damsite in Punjab, the middle Siwalik sandstones are artesian aquifers and several artesian pressure relief wells were provided in the power house area. Limestones, because of the selective dissolution in circulating groundwater, have solution cavities in them. Adequate treatment in terms of grout-filling of the cavities is to be taken in such cases. Schists, phyllites shales, siltstones and clay stones are relatively incompetent and need proper attention. Clay seams, if present, are totally excavated since clay is not only incompetent but also swells on saturation with water. It also increases pore pressure and reduces frictional resistance, thus contributing to sliding. Contacts of igneous intrusives and the host rocks often are fractured and jointed. Further, due to contact metamorphic effects, an alteration zone may also be present on both the contacts of the country rock with the intrusive. In the Ukai damsite in Gujarat, the contacts of basalts and the dolerite dyke were the weak zones. Similarly, at the contacts of a dolerite dyke with the host granite gneisses in the Nagarjunasagar dam area, shear zone with heavily crushed rock was found. The dyke itself was also jointed. Excavation followed by back-filling with grout was adopted.

Exploratory Investigations for Site Location

Phase I

1. All the available information in terms of topographic maps, geological maps, seismo-tectonic information and technical reports is summarized to have a preliminary understanding of the area and also to formulate plan for investigations to obtain a detailed account of the physiographic, geomorphological , geological and tectonic status of the area.
2. Geological mapping of the site and the vicinity for location of the dam and reservoir region is undertaken. This results in the preparation of a detailed geological map incorporating the formation details, boundaries of the formations, strike and dip values of the formations , joints along with the trends and nature of joint surfaces, presence of weak zones with faults, folds as also mapping of weathered regions. Regions of old slides, recent active slides and weathered zones of the formations are also marked on the prepared map. Evidences of neotectonic activity, if present, are recorded and areas marked in the map.
3. The field geological map is integrated with available aerial-photo information and satellite imagery data to produce a composite map. Basic details of both these techniques are presented in detail in the chapter dealing with site investigations. Aerial photo interpretation provides information like lithological variations of the formations, geological structures, texture of unconsolidated formations (whether coarse or fine-grained), different landforms, landslide areas, sink holes and abrupt changes in slope. Drainage patterns of different types reveal the formation characteristics and the structural control of drainage. Satellite data, digitally processed and enhanced provide a regional picture on various features related to geology, geomorphology and physiography of the area under consideration. Remote sensing imageries of an area pertaining to different time periods (temporal or time-based data) enable the monitoring of changes in the area during that time interval. For example, the channel migration (meandering) along with alluvial landforms can be identified. Presence of incised meanders is indicative of neo-tectonic uplift of the area.
4. Geophysical investigations including the shallow seismic refraction study and electrical resistivity surveys are undertaken to obtain continuous information both spatially and to depth. It may be mentioned that while the cross-sections of geological maps can only be approximated to limited depths, geophysical methods provide with certainty an understanding of the formation in terms of the individual thickness values and the respective lithological variations. Elastic parameters (elastic modulii, Poisson's ratio) for each of the individual formations occurring at different depths can be estimated right from the surface using their seismic wave velocities. These are known as the dynamic elastic parameters, which are more accurate being obtained in situ without disturbing the material, in contrast to the laboratory determinations on samples. Tracking of features like faults is also carried out using these methods. Presence of solution cavities (caverns) in the subsurface can not only be located but their geometry in three dimensions established using seismic refraction method. Similarly buried channels along with their depth, pattern and the geometry can be located. Details regarding the methodology and field procedures for geophysical investigations are presented in the chapter on site investigations.

5. Drilling, in vertical and inclined directions, is undertaken at this stage. Drill hole data enable a correct appraisal of the nature of compositional and structural variations in the formations to depth. Panel diagrams along different cross-sections of the area using drill hole logs enable correlation of the formations and structures as also an understanding of the entire subsurface geological picture.
6. Geophysical logs (both resistivity and self-potential) obtained in these drill holes provide significant information on the subsurface geological status in terms of variation of lithology and grain size within individual formations and occurrence of permeable zones at different depths.
7. Field investigations undertaken include also the standard penetration tests, plate-bearing load tests and water pressure tests in bore holes.
8. Groundwater study and preparation of groundwater contour maps are important in pre-construction investigation. Impounding water in the reservoir may modify the groundwater situation. Besides, seepage in the form of springs might be evidenced within the reservoir area. In such a case, seepage details including the nature of formation, source of seepage water, elevation at which the seepage is present and any possible relation to geological structures like fractures or faults are to be recorded. The objective of these investigations is not only to assess the present status of groundwater but also to visualize the impact of reservoir-filling on the foundation and abutment formations.
9. Information on the present channel conditions and their possible modification after the proposed structure starts operating is vital. The erosion status in the catchment area and data on the rate of sediment load transport over the area, collected over a period of several years, would be useful to arrive at a safe estimate of the siltation rate in the proposed reservoir. This would enable working out the design estimate for the life expectancy of the reservoir.

Phase II

10. Once all the field data is obtained from the above approaches and analyzed, locations are identified for samples to be collected for laboratory investigations. These samples are of different categories such as those from drill cores, hand-auger borings and excavated pits. For semi-consolidated and unconsolidated material, both disturbed and undisturbed samples are collected. Laboratory investigations include assessment of properties like strength (both unconfined compression as also triaxial), elastic moduli, permeability, extent of fracturing on the basis of which RQD of the material to be designated and the degree of weathering (weathering index).

    All the data on integration will provide the thickness of overburden, bed rock topography with contours, depth to water table, areas of potential seepage and enable delineation of weak zones that require treatment of one kind or another to make the formations competent.

Foundation treatments

The treatment procedures vary with the specific problem. The weathered zones are stripped. Entire foundation area is grouted with consolidation grouting. Weak zones are drilled to depth and pressure grouting is done filling with grout material to provide soundness to the formation. Grout curtains are provided through placement of grout holes at chosen spacings and often in a radial way to block seepage across the foundations. Grouting of weak formations and modification of slopes along the abutments are also resorted to.

Details regarding drilling methods and foundation reinforcement procedures are discussed in detail in chapter 6. In addition, the Indian examples of selected dams presented in this chapter illustrate specific treatment procedures adopted in different situations.

Seepage Problems in Dams

Seepage through dams, foundations and abutments poses a serious threat to the stability of the dam. Seepage contributes to the pore pressure in the foundation material and abutments. It also causes piping problems. For controlling seepage, proper characterization of the foundations and abutments is important. In addition, potential seepage paths need to be identified.

Treatment for arresting seepage essentially involves blocking the seepage paths. The exact treatment method to be adopted in a situation is essentially site-dependent. Each case has its own characteristic foundation and abutment materials. Following measures are normally adopted:

- Pressure grouting
- Consolidation grouting
- Cut-off trenches backfilled with neat cement
- Chemical grouting
- Grout curtains
- Down-stream drainage system installation

Details on grouting methods are outlined in chapter 6. Pressure grouting is adopted only if the formations are strong and not fractured. On the other hand if the formation is relatively weak and fractured, consolidation grouting is preferred. Grout curtain is also quite popular to provide an effective barrier to seepage. Sometimes, to release excess of pore pressure at a later stage, additional grouting is also adopted to widen and deepen the existing curtains. In karstic areas with carbonate formations having solution cavities, grout filling is resorted to. If the abutment material is extensively jointed and weak in nature, thick concrete walls are placed against the abutments. In all instances, it is extremely important to achieve an intimate contact of the filled material with the adjacent sound rock. For assessing permeability and water intake, effectiveness of grouting operations is assessed by conducting injection tests before and after the grout treatment.

Few Indian examples of seepage failures and the treatment procedures adopted are given below:

In the Kadana dam, a composite dam (earth-cum-masonry dam), located in Gujarat across the Mahi river, considerable leakage varying from 25 to 50 l/min was noticed through the block joints in the drainage gallery. To avert the possibility of choking foundation drainage holes because of the seepage water issuing from the bottom of the gallery joint, redrilling of the drainage holes adjacent to joints has been resorted. Consequently the seepage has reduced.

In the Hirakud dam, an earth dam built in 1958 in Orissa across the Mahanadi river, extensive cracks appeared on the upstream face of the dam. Further, horizontal cracks appeared on the vertical faces of the operational gallery in the right spillway blocks. These cracks were sealed effectively with grouting. Chemical grouting was undertaken for the body of the dam. Drilling of fresh drainage holes in the foundations from the gallery and redrilling of the porous drains in the dam were also resorted (Jauhari, 1999).

Rihand dam, a concrete gravity dam, constructed across the Rihand river in U.P., started operating from 1962. After about 10 years, long cracks were noticed on the upstream side mostly along the lift joints. The cracks, with width varying up to 25 mm were reported to extend to depth varying between 7 and 45 cm. In addition, horizontal cracks appeared in the walls of the foundation gallery. Grouting and other normal procedures were adopted.

In the Nagarjunasagar dam in A.P., leakage through block 44 with discharge varying between 1.5 to 80 l/min was reported. Possible location paths were confirmed to be in the contraction joint between this block and the adjacent one (block 45) on the basis of tracer studies and these have been sealed.

Silting in Reservoirs

The streams flowing through the catchment area into the reservoir carry sediment load, which is deposited in the reservoir. Silting of reservoirs has a serious bearing in that it reduces the life of the reservoirs. For every reservoir, the designed life expectancy is arrived at on the basis of field data obtained from the drainage that contributes water to the reservoir. The life of the reservoir is based on the siltation rate.

Quite often, the assumed siltation rate based on preconstruction investigation is different from the observed rate during the operation of the project. For example, in the case of Tehri dam located in Himalayan terrain in Uttaranchal district of U.P., data for silt load of the catchment stream downstream of the confluence of the rivers Bhagirathi and Bhilanganga are available for a period of 17 years (from 1973 to 1990). The average silt load was estimated as 12.6 ha.m /100km/yr. Taking around 15% as bed load, the net silt rate taken for the design purposes was 14.5 ha.m/100km/yr The dead storage of the reservoir at EL 740 m is reported to be 925 M cu m.

Due to the higher siltation rates observed in the reservoir compared to the rates assumed for its design, the actual life of the reservoir would be lesser than its assumed life. The siltation rates (assumed and observed) as well as the reduction of the assumed life for few selected Indian reservoirs are indicated in Table 8.1.

Table 8.1 Silting in selected reservoirs

Dam	Rate of siltation (ha.m/100sqkm/yr) Assumed	Observed	Percent of assumed life actually available
Bhakra	4.29	5.95	72.2
Tungabhadra	4.29	5.32	78.7
Koyna	6.67	15.24	44
Hirakud	2.52	6.6	38.08
Gandhisagar	3.61	9.64	37.41

(Ref : Jauhari, V.P., 1999 : data source quoted as PAC 1982-83: 103)

Construction of check dams on the upstream side of the reservoir reduces the sediment load carried into the reservoir. Amount of silt can also be reduced in a reservoir through flushing or dredging the accumulated sediment.

Indian Case Studies

In India, out of around 3500 dams operating, the concrete and masonry types constitute around 5% and composite dams around 5%. The rest are all earth dams. Idukki dam is the only arch dam in India, located in Kerala across the Periyar river.

Details of few selected large dams are given below :

Bhakra Nangal Project (Hukku, 1975)

Bhakra dam, a concrete straight gravity dam of 518 m length and 226 m height, is located in Punjab across the river Sutlej. The main works were completed by 1962. The reservoir, known as Govindsagar, has a live capacity of 7771 M cum with a gross capacity of 9868 M cum. Two power houses, the right power house and the left power channel, are located at the toe with five units in each of 90 Mw and 120 Mw respectively. A barrage, located at Nangal, regulates the tail water. The water then goes through a hydel channel (64 km long) with two additional power houses located at Ganguwal and Kotla. The tail water transported through a network of canals, with several branches and distributaries passing through Punjab, Haryana and Rajasthan, is available for irrigation.

The main dam is located the Lower Siwalik formations consisting of sandstone intercalated with siltstone and claystone bands. The formations have a strike along NW-SE strike dipping 70^0 to down stream direction. In all, there are three clay stone bands (Fig. 8.5). The clay stone is the weakest and has a tendency to soften on saturation with water. On the left abutment, one clay stone band of considerable width is present almost at the crest height. The sandstone has prominent shear zones, varying in width up to 18m along the axis as well as at the toe of the dam. On the abutments, a steeply dipping shear zone is also present.

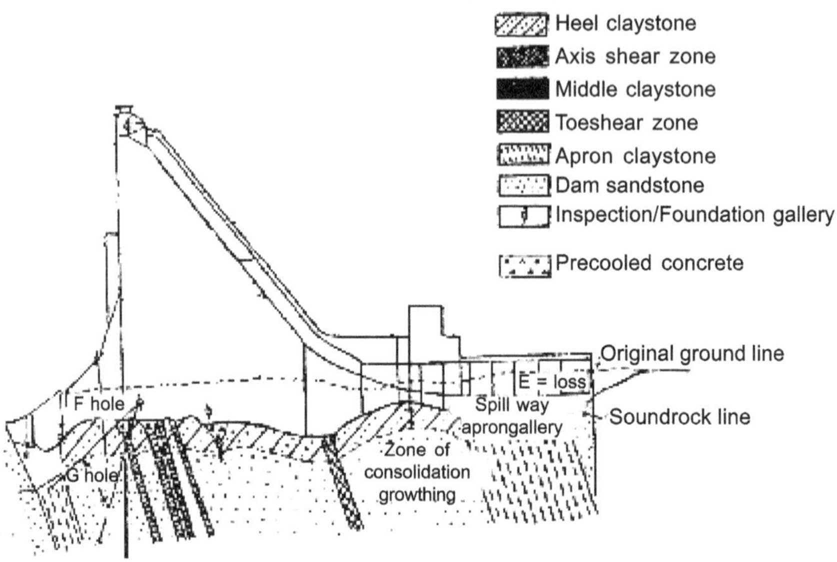

Fig. 8.5 Spillway section of Bhakra dam (Modified from Hukku, 1975).

The following foundation treatment procedures were adopted to reinforce the foundations :

1. Excavation of the clay stone band on the heel side up to 50 m below the riverbed level and backfilling with concrete.
2. Connecting the heel of the dam with the sandstone (below the heel claystone) with a 15m thick concrete slab.
3. Excavation of the spill way apron, tied down to the sandstone overlying the downstream claystone band, to prevent toe erosion.
4. The claystone members and shear zones were excavated up to a depth equivalent to about twice their width and backfilled with pre-cooled concrete.
5. Plugging the cross shear zones on the abutments through concrete backfilling of the of the tunnels to 22 m of depth.
6. Consolidation grouting of the entire foundation area done to a depth of 15 m.
7. Grout curtain provided up to a maximum depth of 90 m provided from the foundation galleries.
8. Grouting tunnels constructed in the abutments extending up to 120 m.
9. To prevent differential settlement of foundation, a cement concrete raft was provided at the left power house located on clay stone (70%) and sandstone (30%).
10. Drainage holes have been provided through the abutments to relieve the uplift pressure.

Instruments were installed for monitoring stresses, strains and joint modifications as well as for recording seismic activity.

During 1961, seepage was noticed through the cracks that appeared on the upstream heel clay stone strut. This was arrested through plugging of these cracks with concrete since grouting was not very effective in stopping the seepage.

Although the design assumption for the silting was 33 M cu m, an increase in the rate to 39 M cu m was reported between 1959 and 1964, which subsequently reduced to around 34 M cu m by 1969.

The Nangal barrage is located in the channel on terrace gravels. Grout curtain was provided extending through the pervious material to control seepage.

Nagarjuna Sagar Dam (Chalapathi Rao, 1972, 1975)

Completed in 1960, across the river Krishna near Nandikonda in Andhra Pradesh, this is a masonry dam (Figs. 8.6 and 8.7) of 124.7 m height and 4865 m in length. The irrigation network is through a right canal (442 km long) and a left canal (320 km long). The power house is located on the left bank. The reservoir has a gross capacity of 11.56 M cum with an effective capacity of around 5.44 M cum.

Fig. 8.6 Nagarjunasagar dam (Photo : G.S. Sukumar).

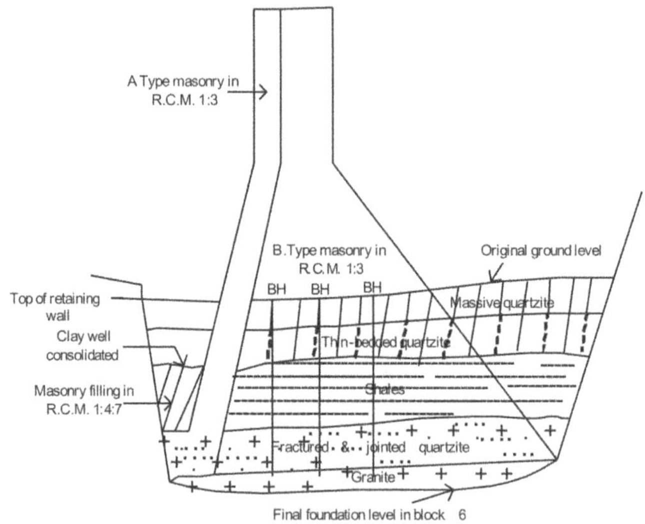

Fig. 8.7 Schematic cross Section indicating geological formations in blocks 5 and 6. (Modified from : Geotechnical features of major dam sites in India, 1982)

The sequence of formations in the project area comprises of the precambrian granite gneiss in the river valley overlain by the quartzites and shales of the Cuddapah supergroup on the abutments with thin layers of soft material. The granite gneiss, coarse-grained, jointed and with vertical foliation trending along NW-SE, is traversed by a dolerite dyke (of 0.9 to 2.4 m width). The dyke is weathered and is associated with a fractured zone of 1.5 m width extending to a depth of about 9 m. The sedimentary formations are near horizontal, with low dips (upto 5^0) towards the downstream.

A low-angle reverse fault, with thickness up to around 5m, is located on the left abutment in the granite gneiss in blocks 7 to 9 extending to a distance of 44 m downstream of the axis.

The foundation treatment consisted of the following :

1. Excavation of the granite gneisses on the hanging wall in the fault zone was carried out with a minimum cover of about 8m of fresh rock retained over the zone.
2. The soft material along the fault zone through 2 to 2.5 m long drifts was excavated and back-filled with concrete. In addition to this, the fault zone was thoroughly grouted through holes from the surface.
3. In addition to grouting in blocks 18 to 23 within the river section containing a dolerite dyke, suitable cut-off shafts at the heel and toe of the dam along the fault zone were provided and back-filled with concrete. To prevent any future settlement, 2 tiers of steel rails were also placed.
4. To prevent sliding and settlement, radial and axial grout curtains were provided in the granite gneisses. The grout curtain holes were placed with increasing radial distances and in four rows.
5. Masonry blankets were constructed covering the sedimentary formations on the hill slopes for a distance of 30.5 m.

6. During excavation, there were roof falls in the horseshoe shaped right tunnel having maximum diameter of 8.23 m (Pasuvemula tunnel) situated in the shale portion.

   These were contained by providing closely-spaced (0.5 to 0.6 m) permanent steel supports along with reinforced cement concrete lining (of 60 cm thickness) of the tunnel portion. Maximum overbreak was found to be around 3 m. In the remaining portion of the tunnel, heavy permanent steel supports were provided with a spacing varying from 0.5 to 1.22 m.
7. Buttress walls, berms and flatter slopes within the jointed rock mass areas were provided to prevent slips along left canal where a number of slips took place. The slips along the right canal were contained by excavating the weathered zone and flattening the slopes.
8. Masonry filling up of the gap between the upstream face of the dam and the excavated slope up to the top of the shale was undertaken.
9. Consolidation grouting all over the sedimentary sequence was carried out. In addition, in the earth dam reaches of the flanks, a grout curtain up to 9 m of depth was provided from a 2 m key trench within the upstream quartzites.

Srisailam Dam (Chalapathi Rao and Narasimham, 1975)

It is a masonry structure (Fig. 8.8) of 143.3m height and 512m length constructed in 1984 across the river Krishna near Srisailam in Andhra Pradesh to generate 770 Mw of power. Three sites were considered – at Niliganga, Pathalaganga and Lingalaguttu for the dam. Finally Pathalganga was chosen as the other sites required either long tunnels to the power house (in case of Niliganga) or a large size dam (as in the case of Lingalaguttu). The reservoir has a gross capacity of 8.7M cu m with an effective capacity of 4.25M cu m.

The foundation geology comprises horizontal quartzites and shales of the Cuddapah supergroup (Fig. 8.9). It has been evidenced that for a span of 3 m between reduced levels 186-189m on the right abutment, the formations are highly weathered, fractured, porous and associated with a 6m wide vertical fault. A thick boulder deposit, with thickness varying between 4 and 28m along the deep course and presence of pools (20 to 30m) both on the upstream as well as the downstream of the dam axis posed serious problems.

Fig. 8.8 Srisailam dam (Photo : M.Raju).

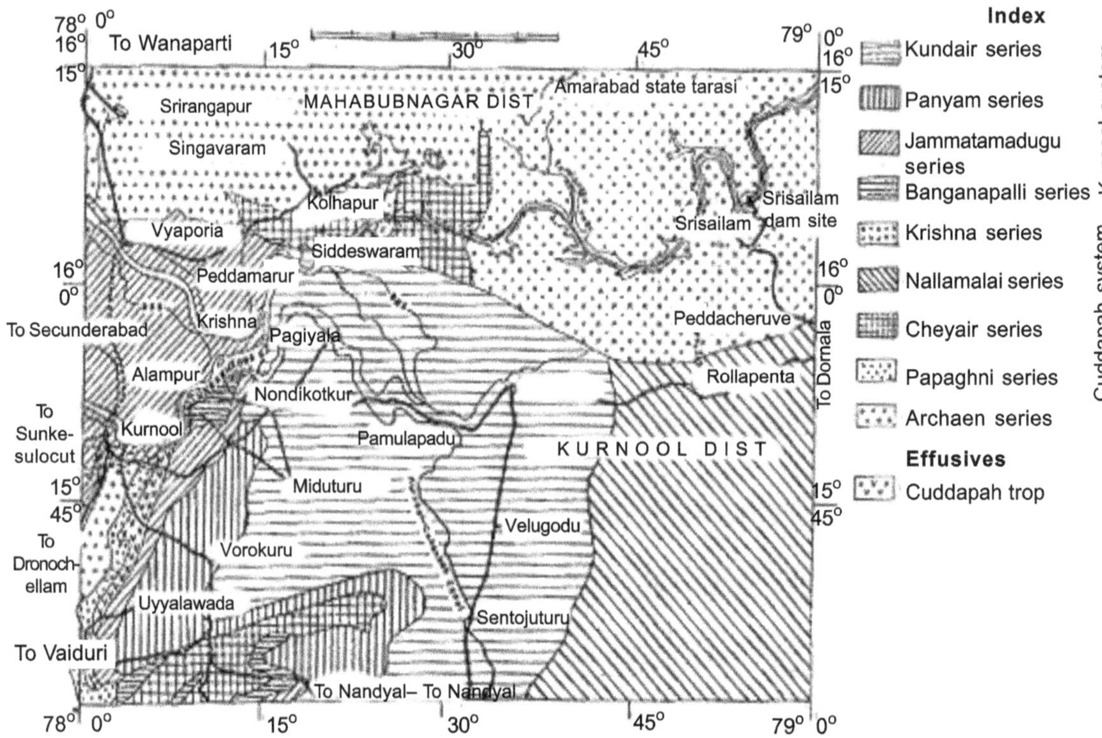

Fig. 8.9 Geological map of the Srisailam reservoir area.
(Chalapthi Rao and Narasimham, 1975)

The foundation treatment involved the following procedures :

1. Two slit trenches of 3 m width were provided near the heel and toe of the dam to confine the soft seams. In addition to grouting the seams, a transverse gallery was provided for drainage and future treatments.
2. Consolidation grouting of the entire foundation area was carried out. Grout holes in a grid pattern were initially placed at 6 m spacing. However, in the toe area, secondary grout holes at 3 m spacing were also placed to achieve sound grouting.
3. To prevent sliding in the downstream area with thin-bedded quartzites having poor shear properties, a highly reinforced cement concrete grillage toe block was provided in the area between 81m from the dam axis and toe which was anchored firmly into the foundation with vertical reinforcement rods up to 10 m of depth.

   In addition, slight upstream arching of the axis of the dam to mobilize the frictional resistance of the shoulder rocks in the deep channel section was achieved through a wedge-shaped excavation tapering towards downstream.

   A single monolithic block, spanning the entire deep channel, was constructed up to 157 m elevation properly resting on the massive quartzites on either side.
4. The weathered rock on the right abutment was excavated up to the end of the dam along the axis through a drift plugged later with concrete. Grouting of the roof and floor portion of the adit was also undertaken to reinforce the sheared zone. Rock chunk of about 6m thickness had to be excavated at the adit face between reduced depths of 476 and 482 m for achieving the monolithic nature of the abutment.

Iduukki dam (Seshagiri, 1975; Roy Mathew, 1996)

This is the only arch dam in India and also the second highest concrete dam in India. Completed in 1974, it is located across Periyar river in Kerala. It is a non-overlowing type. The height of the dam is 167.68m and has a length of 365.85m.The double curvature shape has resulted in the saving of volume of concrete by 60% compared to a gravity dam of equivalent height. The reservoir of 60 sq km with a catchment area of 649 sq km is formed by three dams in the Idukki project of which the Idukki dam is one. The other two dams are - a concrete dam across Cheruthony river and a masonry dam at Kulamavu located upstream of Idukki. The water from the reservoir is transported through a tunnel to the underground power house at Moolamattom. The installed capacity of the power house is 780 Mw with a potential of 230 Mw.

Fig. 8.10 Idukki dam (Courtesy : Hindustan Construction Co., Mumbai).

The rock types in the project area are charnockites with granitic intrusives and biotite-hornblende gneisses of Archaen age, Foliation in general trends in NW-SE direction with very steep dips. On the left abutment, an oval-shaped rock hump (Fig. 8.10) is present separated from the main left abutment by a 15 cm wide shear zone on the downstream side. The dam is asymmetrical and parabolic on the downstream side to take care of this rock hump. To counter this shape, a wing hall (35 m high) has been provided on the upstream side. To avoid the danger of rock falls from this hump which is close to the arch, the rock mass is anchored, through closely spaced rock bolts, into the stable rock. Adequate drainage has also been provided. In the gneiss, the shear zones and clay or gouge–filled jointed horizons were excavated and grouted.

On the right bank, rock falls have taken place due to weak and jointed formations. To prevent their occurrence, 3 rows of 6.09 m long rock bolts were provided at a spacing of 1.52 m. Consolidation grouting of the foundations has been carried out through 824 holes (of 50 mm diameters spaced at 3 m) up to depths varying from 10 to 15 m of depth. No grout curtain was felt necessary, as the foundation material is highly impervious.

Hirakud Dam (Ref : Geotechnical features of major dams in India, 1982)

This is the longest earth dam in the World with the biggest artificial lake in Asia. The dam, completed between 1958 and 1966, is located across the river Mahanadi, about 15 km upstream of Sambalpur in Orissa. This multipurpose composite dam is, with a left earth dam, left concrete dam, right earth dam and a right concrete dam, meant for flood control, irrigation and power generation. The dam has a total length 4800 m (of which around 1148.5 m is of masonry and concrete) with the left and right dykes of 9337 m and 10759 m in length respectively. The reservoir has a gross storage capacity of 7189 M cum with a live storage around 5375 M cum. The project caters to the irrigation of a total area of around 6.99 M ha (including all canal command areas). Two power houses, Hirakud I at Burla and Hirakud II at Chiplima have a total installed capacity of 307.5 Mw. Tail race water from Hirakud I power house to Hirakud II power house is taken through a 25 km long power channel.

Two islands in the river have separated the flow into three channels. The concrete dam portions are constructed on the left and right channels. The project site is composed of metamorphic rocks (of Archaen age) and sedimentary formations (of Cuddapah and Gondwana sequence). The Cuddapah formations are traversed by igneous intrusions.

In the left earth dam area, the contact between the overburden and the gneisses being leaky has been grouted. In the left concrete dam area, the fault zone within the granites as also the weak schists was grouted through consolidation and curtain grouting. In the right concrete dam, the entire foundation area has been traversed by a fault of 150 m width. The rocks in this zone are shattered and also the fault gouge (altered soft material) is present. Both these weaken the formations in this area. The entire fault zone has been excavated up to 15m of depth and consolidation grouting was carried out. As an extra measure, a transverse gallery was provided below the main drainage gallery. In the power house area, after excavating the fault zone to a depth of about 25m, a reinforced concrete raft was provided and the structure constructed on it.

Rihand Dam (Rao, 1975)

This is one of the largest projects undertaken in India and was completed in 1962. The Rihand dam is a straight gravity concrete dam (Fig. 8.11) located near Pipri in Sonebhadra district of U.P. Built across the Rihand (river a tributary of Sone), this dam has a length of 935 m with a height of 92 m. The power house with an installed power capacity of 250 MW was commissioned in 1962. The Reservoir, known as Gobindaballabh Pant Sagar, has a capacity of 10625 M cum^3.

Fig. 8.11 Rihand dam (Courtesy: Hindustan Construction Company, Mumbai).

The rock types in the dam sita area are granites, gneisses, schists , phyllites, amphibolites and quartzites.These belong to the L. Cuddapah series. About 3 km upstream of the dam, Talchir boulder bed and shales (of Gondwana sequence) occur. These are succeeded by the Barakar sandstones and shales (of Gondwana sequence) at the boundary of reservoir.

Based on the exploratory investigations, the dam is designed with a straight alignment in the central portion and with curvature towards the abutments. Investigations from about 200 drill holes have revealed a deep weathered zone (of about 25 m) of the foundation rocks. The entire weathered zone was stripped and backfilled with chilled concrete. A grout curtain up to 30 m in the river bed and 22 m on the abutments was placed through the gallery. In the power house area, amphibolites and mica schists are soft and had to be grouted. The contact between the gneisses of the power house area and the granites of the damsite is sheared. In all, seven prominent shear zones have been identified extending from the axis to the toe of the dam. To avoid any differential settlement, a vertical joint with separation has been provided at the contact zone of the upstream wall of the power plant and the toe of the dam.

Tehri Dam

This is one of the World's largest earth and rock-fill dams (Fig. 8.12). Situated about 90 km from Rishikesh in Uttaranchal, it is across the Bhagirathi river (a tributary of Ganga) about 1.5 km downstream of the river's confluence with Bhilangana. The dam is 260 m high with a base width of 1125m at the riverbed and 20 m at the top. The upstream and downstream slopes are designed as 1:2.5 and 1:2 respectively. The reservoir is planned for a live storage of 2615 Mcum and gross storage of 3540 Mcum. The project when completed will cater to irrigation of 270,000 ha in addition to stabilizing the existing irrigation of 600,000 ha. It will have a final installed capacity of 2000 Mw of power. It is planned to generate power of 1000 Mw from four generators located in the underground power house located on the left abutment.

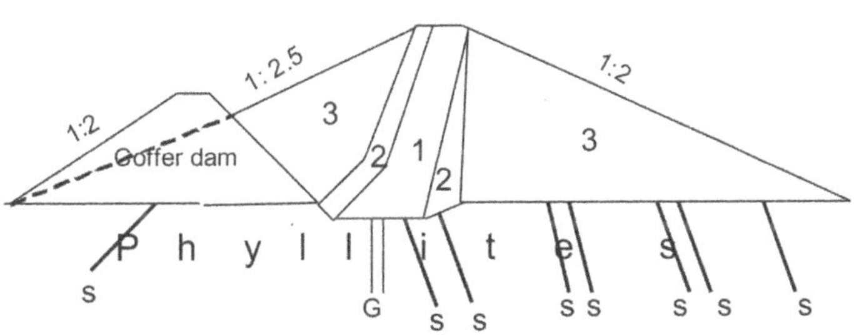

G :Grout curtain S :Shear zone

1 : Well-graded mixture of clay -silt-sand-gravel-boulder

2 : sand and gravel transition 3 : Silt-sand-gravel -boulders mixture

Fig. 8.12 Schematic section of Tehri dam
(Modified from : Geotechnical features of major dams in India)

The dam has been under construction since 1978. The cofferdam has been completed in 1996. Two tunnels have been closed and water let into the reservoir. The final phase of reservoir filling through closing of another two tunnels is being planned. The project is a part of the Tehri Hydropower Development Corporation (THDC), a joint venture of both the Central and State Governments. It has been the most controversial project since its initiation.

The dam is situated in a highly seismic central Himalayan active zone. According to reports, the dam is designed to withstand earthquakes of magnitude up to 7.2 M on Richter's scale. The main controversy is centered on the possible hazards in the event of a severe earthquake in the region exceeding the same. Landslides are very common on the steep slopes of the reservoir. The experience elsewhere (Vaiont dam reservoir in Italy having been flooded due to massive landslide in 1963) has been a cause of concern in this regard. In addition, with the final filling of the reservoir, a vast region including the main township will be affected. It is reported that around 23 villages will be completely submerged with an additional 72 of them partially affected. A new Tehri township has been constructed (about 35 km south of the existing Tehri town) at an altitude of 1550-1900m. The problems of rehabilitation have been attended in providing all associated facilities and financial compensation.

Himalayan rivers including Bhagirathi are snow-fed. The extent of erosion and sediment load in these rivers are quite high because of the high velocities involved. As already indicated earlier, while the erosion is a function of square of the velocity, load- carrying capacity varies as sixth power of the velocity for any stream. Siltation in the Tehri reservoir has been studied (Singha and Gupta, 1982). The erosion is extremely active in the Bhagirathi and Bilangana catchment areas where the slopes are steep with scanty vegetation and the formations are weak and structurally deformed. With heavy silting, the proposed life of 100 years for the reservoir may turn out to be around 40 years (Pradeep Kumar and Ajay Kumar Patel, 1999).

The dam is on Chandpur phyllites (Ordovician-Silurian in age). These formations, dipping in the downstream direction, are intercalated with argillaceous and arenaceous bands. Several shear zones have been identified in the phyllite within the downstream half of the dam. Three important tectonic lineaments are present in the region. A thrust zone (designated as the Srinagar thrust) is about 5 km towards northeast of the dam site (Pradeep Kumar and Ajay Kumar Patel, 1999). Several other faults have also been identified. The main problems envisaged during the exploratory and the design stages were: provision of a good foundation grade for the central area, treatment of the shear zones, leakage through weak zones and possible unequal settlement of the foundation resulting in the cracking of the central core. For an acceptable foundation, the entire weathered zone has been removed below the central area. The area has been grouted for reducing the permeability and strengthening the foundation. To accommodate huge flood discharges, proper chute spillway and multiple silting basins were designed.

9

Engineering Geology in Tunneling Practice

Geological considerations play a significant role in the choice of a right tunnel alignment, selection of proper excavation techniques and for providing appropriate supports for the stability of the tunnel face as the excavation advances. Reconnaissance for the site and tunnel alignment involves several investigations. As the tunnel excavation advances through the access from one side, it is very important to have a proper understanding of the ground so that problems encountered can be minimal.

Tunnels are constructed for several operations:

1. ***In mining practice :*** Adits and shafts for reaching and advancing of the working face and for the transport of workers and material.
2. ***In hydroelectric projects :*** Diversion tunnels for channel diversion and pressure tunnels for power generation.
3. ***For water supply and sewerage disposal :*** Service tunnels
4. ***Transportation :*** Subway rail and road systems; highway transportation
5. ***Underground power houses :*** Large caverns with access tunnels
6. ***For securing rock bolts on the tunnel faces :*** Auxiliary tunnels away from the main tunnel with access to both
7. ***For laying cables and service lines :*** Utility tunnels

Accordingly, the size of a tunnel varies. On the basis of size, tunnels are classified broadly as macro-tunnels (of diameter exceeding 0.75 m) and micro-tunnels (of diameter up to 0.75 m).

Depending on the nature and competency of the ground, tunnels are classified as :

Hard rock tunnels : The tunnel alignment is essentially through competent rock mass with little or no groundwater seepage. The support needed is also limited to the roof essentially. Commonly, the excavation is initiated through blasting.

Soft rock tunnels : The tunnel alignment is through unconsolidated, semi-consolidated or highly weathered material, which has very little standup time. Tunnel supports are needed for the excavated face of the tunnel. Often groundwater problems are encountered in terms of seepage, settlement of wet material or even flowage of material. In soft rocks and unconsolidated material, where groundwater problems are anticipated, tunnel excavation is accomplished under dewatering conditions through sinking of wells and lowering the water table to below the tunnel level. The excavation is advanced in very soft ground by providing a shielding all around and this is shifted as the tunnel excavation advances.

Investigations at the Planning Stage

1. Preparation of a topographic map, if a detailed topographic map is not already available.
2. Preparation of a geological map with all the details covering lithology of the formations, and structural geological details (joints, lineaments, folds and faults). This is carried out through field mapping and also incorporating available information from existing reports, aerial photographs and satellite data.
3. Preparation of groundwater contour map for the region.
4. Subsurface geological (drilling at chosen locations) and continuous geophysical (seismic refraction and electrical resistivity) investigations for establishing the extension of the geological features to the subsurface depths of operation.
5. In situ determinations of permeability, and also the elastic parameters.
6. Preparation of longitudinal sections along proposed tunnel alignments to visualize the extent of each formation along the tunnel length, weak zones likely to be encountered and possible stability problems.
7. On the basis of the thickness of the material overlying the proposed tunnel excavation, initial stress conditions to be calculated all along the proposed tunnel alignment.

Laboratory Investigations

8. Strength and deformation tests on the rock (both under dry and saturated conditions), which enable the engineering classification of rocks.
9. Geotechnical characterization of soils.

   Based on the analysis of the data, the best possible alignment of the tunnel is decided from among the various alternatives considered. Once the alignment is decided, the tunnel specifications including the supports are finalized. This is followed by the choice of equipment and construction of the tunnel.

During construction, if there are problems that arise due to unanticipated geological situation or sudden problems owing to the stress adjustment and consequent ground behaviour, midstream corrections are done like reinforcement of the tunnel walls and roof. If the problems encountered are complex, the alignment of the tunnel for the problem span is deviated through relatively competent and safer path based on geological considerations.

Geological Considerations

1. *Lithology of Rock Formations*

Lithology or the composition of the rocks is an important parameter influencing their performance. Among the igneous, sedimentary and metamorphic rocks, igneous rocks are in general compact and competent. Unless these are highly jointed or weathered, normally no lining is required for the tunnels designed in the same. Volcanic igneous rocks, being often vesicular, are porous and permeable. This coupled with extensive jointing poses a threat of water seepage in the tunnel.

Among the metamorphic rocks, quartzites and gneisses are massive and competent. Phyllites and schists are problematic due to the presence of foliation (planar structure) and presence of susceptible minerals like mica and clay. Performance of the slates depends on the orientation of the slaty cleavage to the tunnel alignment. Carbonate minerals (calcite and dolomite), if present, in the slates go into solution in presence of water with time and consequently impairing the long-time performance of the formations.

Sedimentary rocks in general are less competent. Among these rock types, performance of the sandstones with siliceous (silica) cement is the most preferred one in terms of competency. If the sandstones have carbonate or iron oxide cement or are porous due to poor cementation, the tunnel lining needs reinforcement. Limestones are prone to selective dissolution and hence seepage problems are expected. Shales are the least competent because of the clay content. In addition to their poor strength, properties like slaking and swelling create problems during the construction stage or later. Clay layers are troublesome as they have low strength, swell in presence of water and also undergo creep (time-dependent deformation).

2. *Geological Structures*

The engineering behaviour of rock formations is not only influenced by the lithology, but also in a dominant way by their structural features. The sum total of these aspects is reflected in the formation performance. Hence, the engineering classification deals with the rocks as rock masses, in which the rocks with their associated features such as the weak zones and the status of water saturation are considered together. As already stated in the chapter on engineering classification, the rock mass properties at the site (in situ) are more relevant than the same properties as determined in the laboratory on samples with geometrical constraints and devoid of weak zones.

Tunnel Alignments as Related to Geological Structures

1. *Strike and Dip Orientation*

If the tunnel alignment coincides with the strike of the formations or very close to the strike direction, the same formation continues until the alignment deviates from this direction. This is acceptable if the formations are competent but in the case of less competent formations, it is desirable to have the shortest span of the tunnel within them. Hence it is preferable to have the alignment across the strike in a geological sequence with less competent formations.

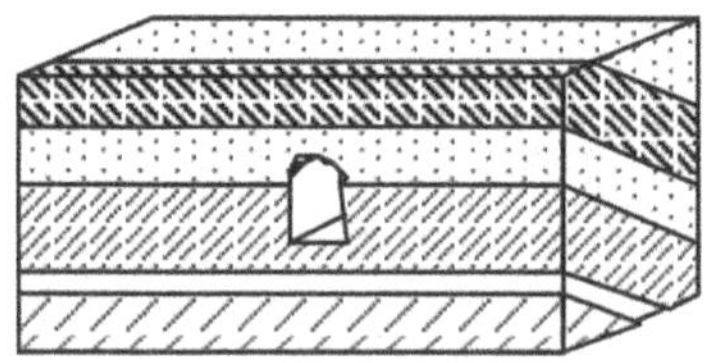

Fig. 9.1 Tunnel through dipping formations.

If the formations have very low dips (corresponding to near horizontal situation), problems are minimum in thick formations as the tunnel intercepts a single formation. If the particular formation is competent, then the necessity of lining would be minimal or not there at all. In the case of dipping formations intercepted by a tunnel (Fig. 9.1), while competent formations do not pose problems, incompetent ones create problems of slippage along the dip directions or seepage into the tunnel.

In the case of jointed formations with vertical dips, lining of the tunnel roof is needed to keep in place the layers from slipping down (Fig. 9.2).

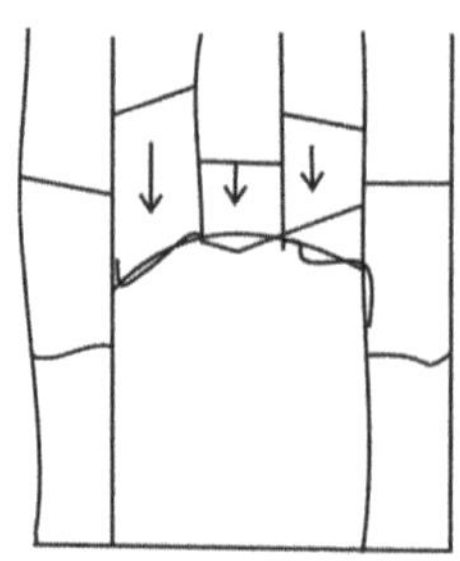

Fig. 9.2 Tunnel in jointed rocks with vertical dips.

2. *Tunnels in Folded Rocks*

Alignment of tunnels in synclinal or anticlinal areas has characteristic problems. In crest region of an anticline, the tensile stresses result in the development of joints or fractures leading to the formation of a weak zone. In the core of an anticline, the oldest formation of the sequence occurs. Similarly, the younger formation of the sequence is located at the core of a syncline. If the tunnel is aligned in the direction of the axis of the axial plane of the fold, the same formation (depending on what the tunnel intercepts) continues.

Thus less competent material can be avoided by locating the tunnel in such a way that it is situated on the limb of a fold and at the level where the problem formation can be avoided. It is advisable to have the tunnel located on the limbs than at the core if possible.

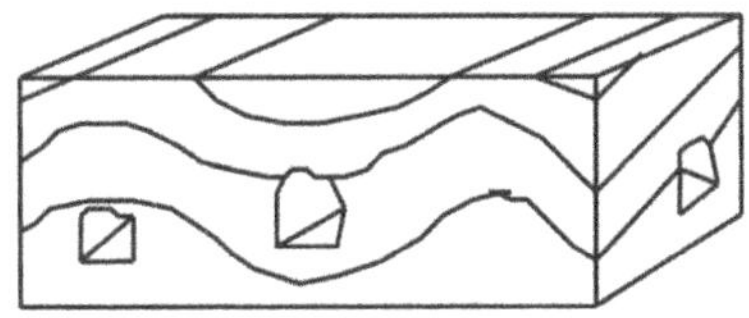

Fig. 9.3 Tunnels in folded rocks.

On the other hand, if the axial plane is at an inclination or perpendicular to the tunnel alignment (Fig. 9.3), repetition of the formations occurs within the tunnel. The tunnel can be so located as to pass through minimum thickness of a problematic formation.

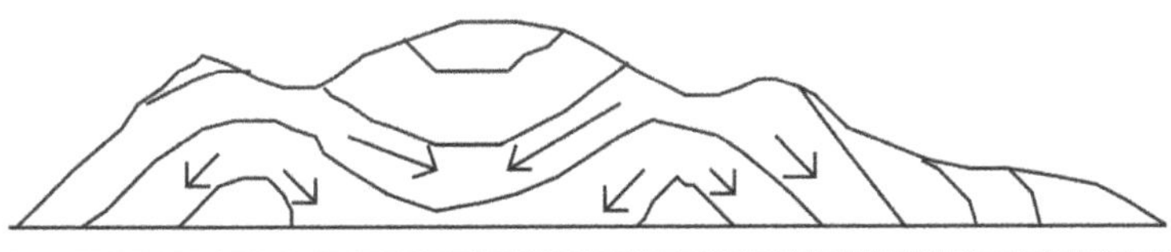

Fig. 9.4 Tunnel perpendicular to strike of formations (Arrows indicate groundwater flow).

Problems of groundwater are to be anticipated in the cases where the formations dip towards the tunnel opening, which is the case with a synclinal structure (Fig. 9.4).

3. *Tunnels in Faulted Formations*

Close to the fault plane, formations exhibit fractures and shattering, resulting in fault breccia. In addition, due to the local heat generated due to the slipping along the fault plane, the formations are burnt or altered into clay-like material known as mylonite or fault gouge. A major fault is normally associated with minor faults and extensive weak zone. Both these aspects contribute to a drastic decrease in the competency of the formations. In addition, the impervious mylonite zones act as barriers to groundwater flow and consequently the groundwater is stored at its boundary. On excavation, a sudden gush of water takes place in such a situation. Fault zones are normally avoided along tunnel alignments. They introduce non-uniform pressures on the tunnel.

However, if they cannot be avoided, the fault zone has to be extensively treated with grout and a strong lining has to be provided against these zones along the face of the tunnel opening.

Faulting, because of the displacement of the formations, suddenly introduces formations of different lithology along the tunnel route. Detailed subsurface data collected during the initial reconnaissance stage, helps in the projection of the faults occurring on the ground surface to depth at the tunnel level. Proper analysis of the data thus would minimize problems during tunneling.

Problems are severe if the tunnel alignment coincides with the strike of the fault. In such a case, the fault plane and the weakness zone are encountered for quite some length along the tunnel excavation. If the tunnel is located in the footwall of a fault, the fault is intercepted towards the roof portion of the tunnel (Fig. 9.5 A) resulting in the instability of the roof portion. It has to be reinforced accordingly.

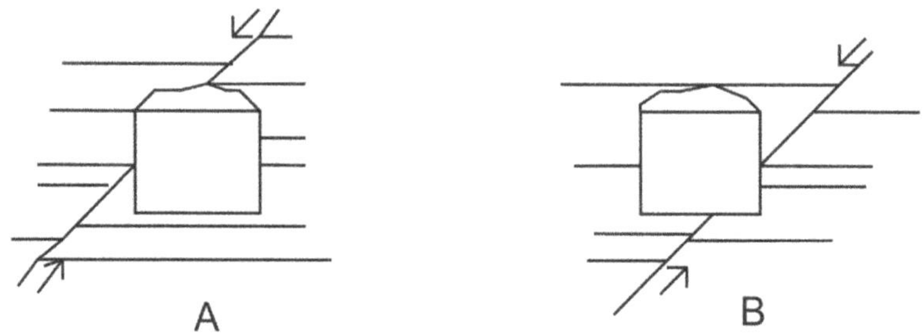

Fig. 9.5 Tunnels in faulted rocks (A: on foot wall side, B: on hanging wall side).

4. *Tunnels in Jointed Formations*

Joints contribute to the instability of the excavation in a substantial manner. In formations with steeply dipping joints, rockfall into the tunnel takes place all around the face, including the roof. Overbreak is greater in the excavation of jointed formations.

Excavation through Blasting

The excavation is to be as close to the tunnel geometry. It is not possible to obtain the exact shape of the excavation conforming to the tunnel specification. There will always be an extra breaking of material, which is known as overbreak. The extent of the overbreak depends on the type of formation and the spacing of discontinuities. Discontinuities or joints play an important role in the overbreak. Considering the possibility of overbreak, a pay line is specified in contracting agreements. The extent of pay line varies with each ground situation. A pay line is defined as the specified limit up to which the payment is made for the excavation of the tunnel considering the minimum overbreak. Overbreak and pay line are indicated in Fig. 9.6.

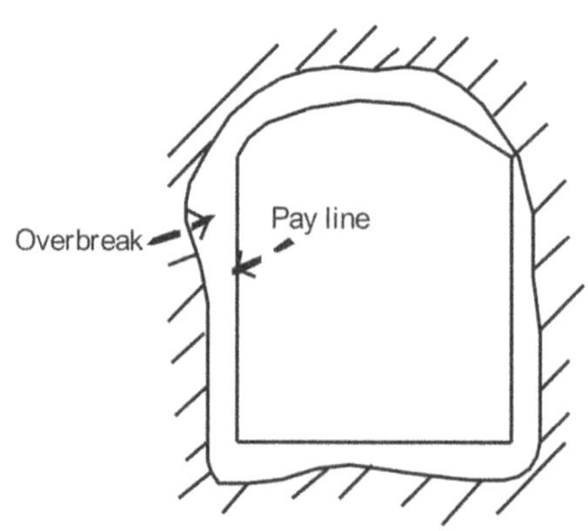

Fig. 9.6 Overbreak in tunnel excavation

Blasting methods : Blasting operations are carried out normally in a sequential manner to minimize the overbreak. Various patterns adopted for blasting are:

1. ***Parallel hole cut (Fig. 9.7) :*** Normally adopted for smaller tunnels. Blast holes are placed parallel to each other and normal to the face. A relief hole of a larger dimension is also drilled but no charge is placed in it. This hole is meant to act as a free face. A modified version of this type has several relief holes located in between the blast holes.

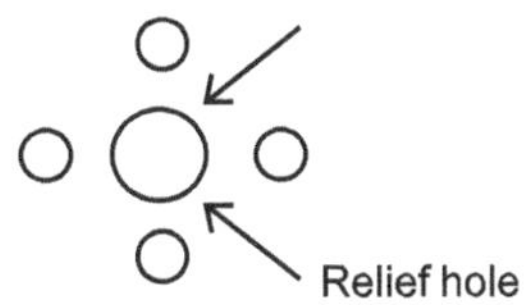

Fig. 9.7 Parallel cut.

2. ***Fan cut :*** The drill holes are placed as in the Fig. 9.8. Once the charge is detonated, more face is available for the advance of the excavation.The sequencing of blasts are also indicated in Fig. 9.8.

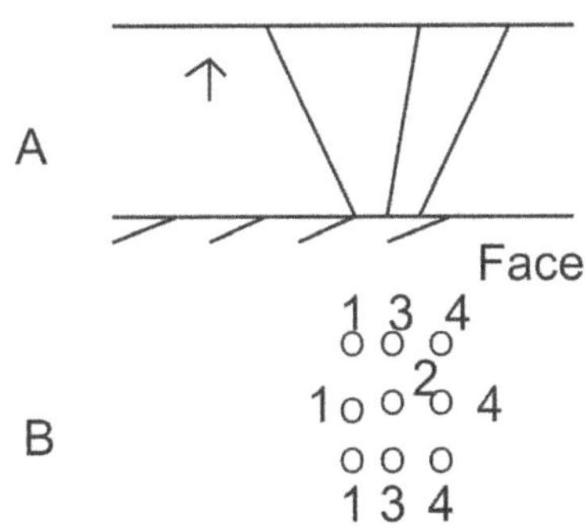

Fig. 9.8 Fan cut : A. Plan, B. Face view (Numbers indicate blasting sequence).

3. ***Smooth blasting :*** For the final shape, small holes are placed also along the circumference and blasting carried out with blasting charge varying at each stage to control the overbreak (Fig. 9.9).

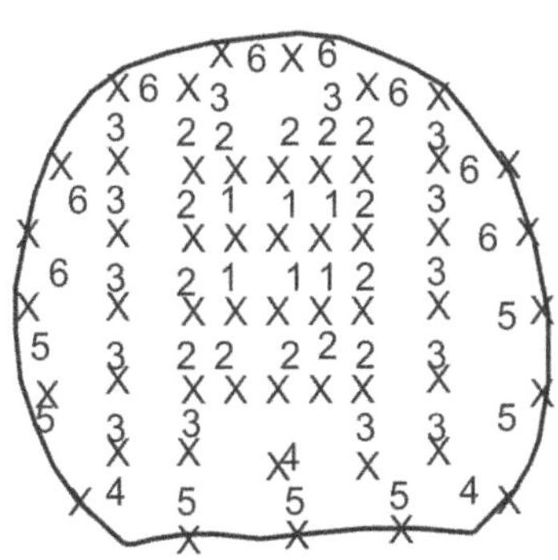

Fig. 9.9 Sequential blasting in tunneling.

4. Blasting and excavation for a cavity (Fig. 9.10).

   Sequence of excavation for surge shaft tunnel is indicated in Fig. 9.10.

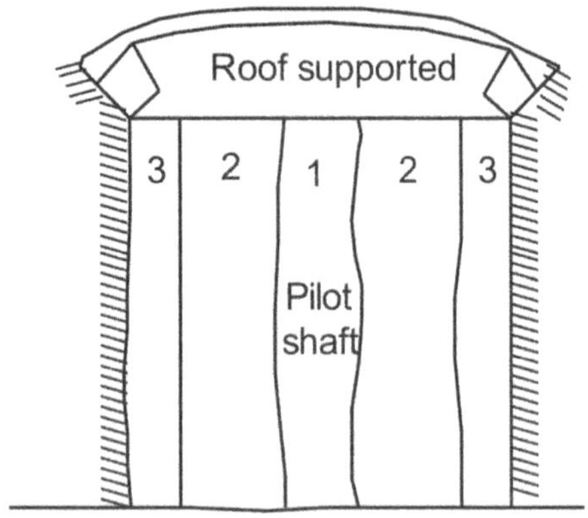

Fig. 9.10 Typical excavation of a shaft (Numbers indicate excavation sequence).

Geometrical Shapes of Tunnels

Depending on the purpose, particular shape of a tunnel is chosen. For example, for diversion of water at dam sites and also for pressure generation of inlet water to the generators, the circular shape is adopted. Circular shapes facilitate avoiding stress concentrations in corners, thus preventing progressive failure mechanisms to operate. For other purposes, the roof is of a circular segment with the side faces kept vertical. The circular arch for the roof is the most stable one. There are other shapes like the horseshoe type. These are common in old tunnel excavations.

Stress distribution during excavation

The rock mass is under confining conditions in the subsurface. As explained already, the overburden pressure at any point of depth can be estimated. Once the tunneling is under progress with the removal of material during excavation, a stress gradient is created due to stress readjustments. The roof and the walls are subjected to stresses. Rocks fail under tension. This is common at the roof portion. Rocks withstand compression and hence the walls of a tunnel will not ordinarily fail except at considerable depth (King, 1996)

A generalized picture of the stress distribution (Terzaghi, 1946) is indicated in Fig. 9.11. The loosened rock portion (KLMN) tends to come down and is resisted by friction along the surfaces KM and LN. In this process, a significant proportion of the total overburden pressure is exerted against the roof and walls of the tunnel. To counteract this, tunnel lining and supports are provided. Primary tunnel linings are normally chosen as thin flexible ones so that the bending moments are minimized and the stress redistribution is achieved without any sectional forces on the lining. Additional supports are given in the form of rock bolts and anchors. All these types of supports are described in detail in the chapter 6.

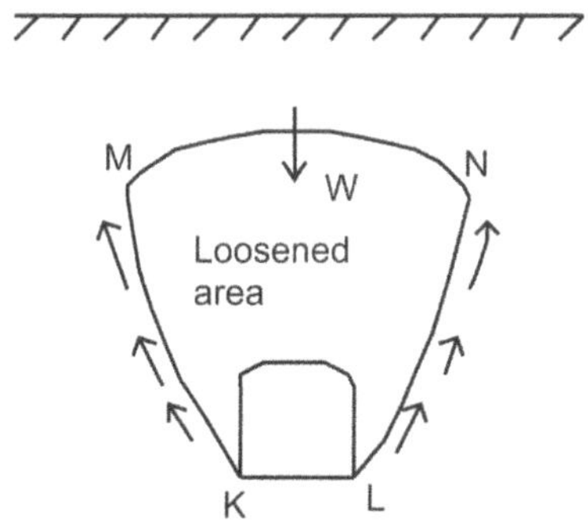

Fig. 9.11 Stress distribution around an excavation.

Ground Failure in Tunnels

Failures in tunnels are of different kinds. These may take place during excavation or after tunneling. The failures may be sudden or of a continuing type. Different failures are as follows:

Sudden failures with less violence :

1. ***Spalling :*** Dislodging in jointed formations
2. ***Bending :*** Deformation resulting in fissures
3. ***Bumping ground :*** Rock displacement and dislodging

Sudden failure with enough violence :

4. ***Rock Bursts :*** These occur at great depths with enormous overburden pressure. A sudden release of strain energy is involved. The maximum strain energy according to Obert and Duvall, (1967) is S_E = S/2E where S_E is the maximum strain energy per unit volume, S is the uniaxial compression stress and E is the Young's modulus. Rock bursts take place in brittle rocks only since enough accumulation of strain energy is possible. In India, these are common in Kolar Gold Mines.

Slow failures :

5. ***Swelling Ground :*** This is the result of change of volume due to expansive minerals (like montmorillonite) swelling in the presence of water. Slow failure is also common in over- consolidated clays.
6. ***Squeezing Ground :*** This is due to slow movement into the tunnel, common in soft clays and highly weathered rocks.
7. ***Flowing Ground :*** Ground movement is similar to that of a viscous material.

Tunnel Supports

Supports are used for keeping the tunnel walls and the roof in place. Several support alternatives are available for use in tunnels. The choice of a particular type is based on the nature of the ground and the type of failure anticipated.

Types of Supports

1. Gunite or shotcrete
2. Rock bolts
3. Wire mesh (welded grid type and woven fabric type)
4. Concrete lining
5. Pre-stressed anchor cables or perfo-bolts
6. Steel ribs or arches

Some of these types are used in combination. For example shotcrete is used to hold a shattered rock in place before the use of rock bolts. All these supports are dealt in detail in chapter 6.

Use of Rock Mass Classifications in the Choice of Tunnel Support

The classifications proposed for use in tunneling practice for the choice of supports incorporate parameters including the unconfined strength, modulus of elasticity, different aspects of the discontinuities, status of weathering and groundwater situation. A brief review of the classifications is presented below:

1. ***Rock Quality Designation (RQD) (Deere, 1968)*** This classification forms an integral component of rock mass characterization schemes of later workers. This method deals with the ratio of cumulative length of rock pieces of 9cms and more of length expressed as a percentage of the total length of the rock core. On the basis of various ranges of RQD values, the rock mass is classified into different categories ranging from very good to very poor type. Details are given in chapter 2.

2. ***Competence factor (Muir-Wodd, 1972)***

   Competence factor, F_c = UCS / net pressure of the overburden.

   Values of Fc are used for tunnel lining decisions:

   $F_c < 2$: Continuous support is required

   F_c 2-10 : Stability is a function the ground situation. Accordingly the supports are decided

   $F_c > 10$: No supports needed ordinarily

3. ***Rock Structure Rating (RSR)***

   As already indicated in chapter 2, three parameters related to geological situation are involved in this classification

   Parameter A : Rock structure features

   Parameter B : Joint pattern and drive direction

   Parameter C : Groundwater flow evaluation

   Ratings for each of these parameters are obtained from the rating tables (chapter 2) and RSR value for a formation corresponds to the sum of all these ratings. It varies from 25 to 100. Formations with RSR less than 27 definitely require supports. For RSR grater than 77, normally no supports are required. In between these two limits, supports are to be provided based on the specific situations.

4. *Rock Mass Rating (RMR) System (Bieniawski, 1976)*

RMR involves 6 parameters (either measured in the field or on samples), as outlined in chapter 2, involving RQD, UCS, joint status (spacing, conditions as also their strike and dip) and water saturation. These parameters are rated for the rock mass under consideration. For this purpose the values in the tables (chapter 2) are chosen. Rock masses are classified based on the total ratings of all the six parameters.

5. *Rock Mass Quality (Q-System) (Barton et al, 1974)*

The expression for the estimation of the quality index, Q, is presented in chapter 2. This involves size of the jointed rock mass (in terms of RQD and number of joint sets), shear strength in between the blocks (in terms of joint roughness and joint alteration) and the active stresses (as evaluated from the joint water status and stress reduction factor). While RQD is obtained from the drill cores, the rest are from field observations. The individual rating tables are indicated in chapter 2.

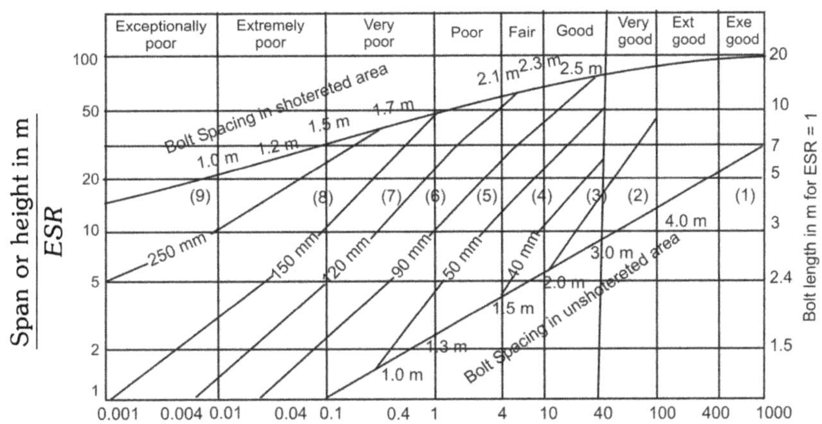

Rock mass quality $Q = \frac{RQD}{Jn} \times \frac{Jr}{Ja} \times \frac{Jw}{SRF}$

Reinforcement categories : 1. Unsupported, 2. Spot bolting, 3. Systematic bolting, 4. Systematic bolting with 40-100 mm, 5. Fiber-reinforced shotcrete 50-90mm and bolting, 6 Fiber reinforced shotcrete 90-120 mm and bolting, 7. Fiber reinforced shotcrete 120-150 mm and bolting, 8. Fiber reinforced shotcrete >150mm with reinforced ribs of shotcrete and bolting, 9. Cast concrete lining

Fig. 9.12 Estimation of support categories based on Tunnelling Quality Index (Q). (Grimstad and Barton, 1993)

The value of 'Q' is used for tunnel support requirements. For this purpose, a chart relating "Q" with another parameter designated as equivalent dimension (D_e) has been proposed by Barton and coworkers.

The equivalent dimension = (excavation dimension) / (excavation support ratio)
(D_e) (span or height or diameter) (ESR)

The excavation dimension and excavation support ratio are to be known for calculating the equivalent dimension.

The excavation support ratios for different excavation categories, as suggested by Barton and coworkers, are given below :

	Excavation category	ESR
A.	Temporary mine openings	3-5
B.	Permanent mine openings (water tunnels for hydro-power, pilot tunnels , drifts and headings for large excavations)	1.6
C.	Storage rooms, water-treatment plants, minor rail and road tunnels, surge chambers, access tunnels	1.3
D.	Power stations, major rail and road tunnels, civil defence chambers, portal intersections	1.0
E.	Underground nuclear power stations, rail stations , public utilities and factories	0.8

Using ESR and Q values, the length of rock bolts (L) and maximum unsupported span are obtained from expressions given by Barton and coworkers (1980).

$$\text{Length of rock bolt (L)} = (2 + 0.15\ B)\ /\ ESR$$

$$\text{Maximum unsupported span} = 2ESR\ Q^{0.4}$$

From the chart of 'Q' versus 'D_e', the appropriate bolt dimensions and spacings can be obtained. The original chart of Barton and coworkers (1974) has been modified by Grimstad and Barton (1993) by incorporating the fiber-reinforced shotcrete requirements (Fig. 9.12).

New Austrian Tunnelling Method (NATM)

This method has evolved through considerable tunneling experience in Alps. This is extremely suited to complex geological conditions. The formations form a part of the supporting system. The formations and the soil are considered as one integrated system. Early temporary support is given through shotcrete immediately after the excavation to limit ground deformation. Shotcrete imparts complete strata support irrespective of the surface irregularities on the tunnel surfaces. Thus temporary supports constitute the main load-bearing means in this method.

For the choice of the support systems, the rock mass is classified into 8 classes as follows :

A1	Stable ground	A2	Slightly over breaking
B1	Friable	B2	Heavily friable
C1	Pressure-exerting	C2	Heavily pressure exerting
L1	Loose ground, highly cohesive	L2	Loose ground, low cohesive

While for A1 class ground no supports are needed, the rest need supports in the form of shotcrete together with rock bolting, forepoling and steel ribs. As the ground condition deteriorates (A2 to L2), these supports are provided at closer spacings.

Selected Indian Examples

1. *Underground Power House in Sardar Sarovar Project (Gujarat)*

This multi-purpose project is located near Navagam (in Baruch District). The dam, of concrete gravity type, is 1.2 km long and 129 m high with a live storage of 5800 Mm^3 of water in the reservoir. There are two powerhouses – an underground one on the right bank (23 m wide, 58 m high and 212 m long) and a surface one (Canal Head Power House). From Navagam reservoir, a lined canal constructed on the right bank, is 120 km long up to the Rajasthan border.

Blasting for the powerhouse cavity was carried out with multiple drift method. The sequence adopted is indicated in the Fig. 9.13.

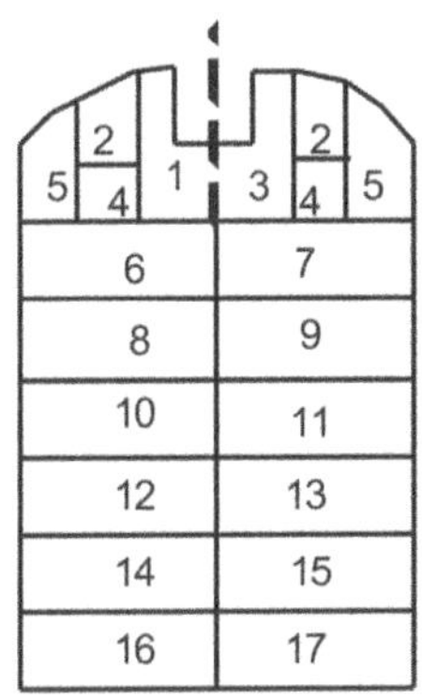

Fig. 9.13 Sequence of excavation of the powerhouse cavity (Menon et al, 1998).

The bed rock within the cavern is composed of basalt containing pockets of agglomerates with three dolerite intrusions. The main dyke, with a trend of 65^0SE, is on the southern end of the cavern. The second dyke is beyond the northern border of the cavern. A horizontal dolerite sill occurs close to the turbine level (Menon et al, 1998). The contact zone of the basalt with the dyke is sheared at the south side. Besides, several shear zones, varying in width up to 0.4 m and few faults are located in the powerhouse cavity area.

The RMR values for basalts ranged from 67 to 72 except in sheared zones (with RMR around 23) and basalts with agglomerates (with RMR around 28).

For the roof arch, tensioned rock bolts (6 m long, 25 mm diameter) along with expansion type shell anchors were provided at 0.87 m spacing (staggered). Also 78mm thick shotcrete in two layers with welded wire mesh (100 mm by 100 mm) in between was also provided in the area with the rock bolts (Menon et al, 1998). Similar support system was given in the shear zone regions but with shotcrete (up to 300 mm of thickness). However, in the dolerite dyke areas, the rock bolts were at a spacing of 1.75 m and with 76 mm thick shotcrete. On the cavity sidewalls, similar rock bolt (6 m long) supports were given except in the middle one-third portion where 7.5 m long bolts were provided.

2. *Tunnels in Jammu-Kashmir Highway Region (Hukku, Chaturvedi and Ashraf, 1971)*

For the fractured areas encountered during excavation, a detailed review of the earlier support system was undertaken based on which the length of the rock bolts were changed to 10 m and 12 m (in place of 6 m and 7.5 m bolts) sizes. Similarly in the shear zone areas, additional reinforcement included inclined cable anchors (25 m long and 50 tonne capacity).

Four tunnels were constructed in the region, Two of them (Nandini and Banihal tunnels) are for mass transportation while the other two (Chineni and U.Sindh tunnels) are for transportation of water.

The general geological set up in the region consists of formations of differing age groups separated by thrust faults. The general sequence has older formations thrust over the younger ones due to these thrust faults. In the South, along the Main Boundary Fault, the L.Siwaliks (of U.Tertiary age) consisting of sandstones, claystones and boulder beds are overlain by the older Murree formations (of Oligocene –Miocene) composed of alternate layers of sandstones, siltstones and claystones.The Murree Thrust separates the Murree formations from the older Carboniferous group represented by slates, shales and limestones. In the North, the Panjal Thrust separates Carboniferous formations from the overlying older (Archaen) formations (shales and slates).

Details regarding the geological considerations for the alignment of these tunnels and the support systems provided are reported by Hukku, Chaturvedi and Ashraf (1971).

The Nandini tunnel, located at about 27 km from Jammu, passes through the L. Siwalik formations. This tunnel has two parallel traffic passages (around 120 m long) for a 2-way traffic. This tunnel system is of horse-shoe shape and passes through the southern limb of an anticlinal fold. The alignment is across the strike of the formations. This is also well suited from geological considerations. No problems of serious nature were encountered. Irrespective of this, steel supports at close spacing were provided over its entire length.

The Chineni tunnel group has six tunnels of which the longest is of 6.5 km. The tunnels are of horseshoe shape. They pass through the Murree formations. Their alignment across the strike of the formations was not possible because of the topography. The first tunnel (1A) is curved in its alignment to avoid a jointed zone. The second tunnel (1B) was so aligned that it was outside a slide zone. But to achieve a short length, a portion of it had to pass through talus material. Closely-spaced steel supports were provided in talus zone and in the highly disturbed areas. Heavy seepage of water has been reported to be responsible for the bending of steel ribs at places. So the spacing of the steel ribs has been reduced from 0.9 m to 0.45 m. The next four tunnels (2A, 2B, 3 and 4) have bends in their alignments. For tunnel 2A, a bend of about 156° has been provided close to the inlet side. For tunnel 2B, two bends had to be provided. Even then, a portion of the tunnel had to be along the strike of the formations. Excessive overbreak and collapse were experienced in this section. Alignment for tunnels 3 and 4 was at an angle to the strike. Cement-concrete lining (minimum of 38 cm thickness) with steel reinforcements was provided.

The Banihal (Jawahar) tunnel is of horseshoe shape and passes through the U. Carboniferous formations. This tunnel, of length around 2540m, has two passages (for traffic). The igneous rocks (Panjal traps) are mostly unlined. The slate section is lined.

This tunnel had to be located around 200 m below the elevation level of the Banihal Pass to avoid problems of snow in winter. The final alignments, chosen from four possible alternatives, were based on the consideration that much of it should be within the self-supporting trap rocks. Heavy seepage (up to about 6800 l/min) was experienced in the limestone portion. Waterproof insulation, concrete lining and longitudinal drains were provided.

The U.Sindh tunnel, a lined tunnel of around 240m in length, is of D-shape and passes through the slates. Because of an unstable zone, a bend in its alignment was given. Maximum angle (around 118^0) to the strike of the formations had to be provided to avoid excessive overbreak. Steel supports were provided only for about 30% of its length. For economic reasons, gunniting was done instead of shotcreting.

3. *Excavations and Support System for the Underground Power House, Yamuna Hydel Project : (Kadkade, 1971)*

The underground powerhouse of the Yamuna Hydel Scheme at Chibbro in Dehradun district of Uttaranchal is situated within soft Siwalik formations. Water from the Ichari dam, located on the Tons River (a tributary of Yamuna), is diverted into this powerhouse for the generation of power. This was the first underground powerhouse constructed in the Himalayan region The tunnel connecting the intake area at the dam to the powerhouse is located in the Jaunsar synclinal region composed of quartzites, slates and limestones. The rocks are highly jointed and severely folded. The powerhouse cavity, located in limestone with several shear zones filled with gouge material, extensive jointing and slip planes, posed enough challenges for its construction. The sequence of excavation of the cavity is indicated in Fig. 9.14. To control overbreak, blasting charges for each sequencing set were varied in quantity.

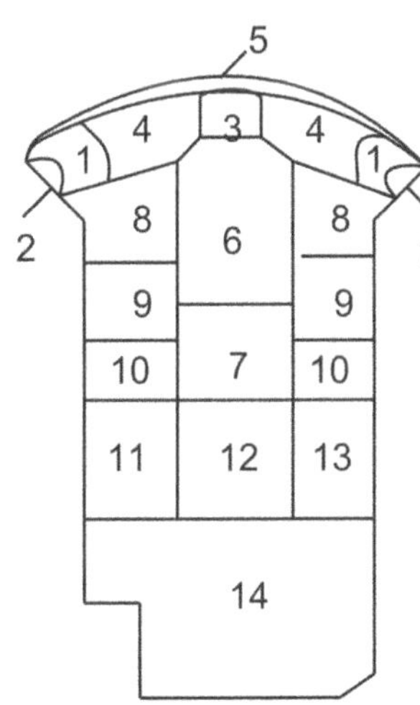

Fig. 9.14 Sequence in excavation of the Chibbro power.

The highly deformed nature of the rock required simultaneous support installation along with the advance of the construction work. The crown of the powerhouse cavity was first supported in segments through steel arch supports. For supporting the walls of the powerhouse cavity, prestressed steel cable anchors (perfobolts) were provided. The walls are anchored to the competent rock in which the support tunnels (auxiliary tunnels) are situated on either side of the main tunnel.

Details of these prestressed anchors (known as the Glifford Udall CCL prestressing system)

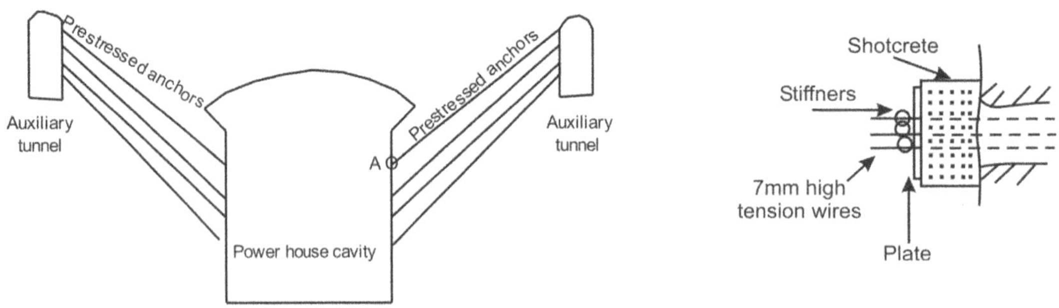

Fig. 9.15 Sketch indicating prestressed cable Details at 'A'; Support system (Modified after Kadkade, 1971).

are indicated in Fig. 9.15. Pre-grouting of the holes drilled for the passage of the anchors was carried out and the hole was drilled again at the desired angle up to the auxiliary tunnel. Stressing of the cables was done after their installation. The cables were prestressed after a period of time and then the hole was grouted finally.

10

Geological Hazards

Earthquakes, volcanoes and landslides pose a threat to the Earth's stability. They cause considerable destruction in many ways. During an earthquake, the ground motion results in damaging buildings, dams, dislocating roads, railway tracks, alter the course of surface and groundwater flow, and opening up of ground fissures. Volcanic activity brings devastation on a large scale. Not only the whole landscape changes, but even the climatic changes are brought about due to profuse emission of a very dense cloud of gases. Thick forests are denuded, vast areas are buried under thick volcanic debris and parts of the volcanic cone are thrown out due to intense fractures. Landslides and slips are common phenomena during earthquakes. Huge flow of pyroclastic debris mixed with water flows along with the ground material during slides causing damage to whatever lies in its path. These flows, called lahars, fill valleys with thick deposits of their load, block channel ways and destroy structures en route their flow as they carry huge assorted load of varying dimensions. Sudden outburst of volcanoes triggers earthquakes. Thus earthquakes, volcanic activity and landslides are all inter-related. Landslides are also common in hilly terrain with unstable hill slopes. Mapping of the hazard zones is an important activity in civil engineering construction practice.

Earthquakes and volcanoes are related to crustal instability. The earth's crust is broken into 13 major plates (Fig. 10.1), which are in constant movement (1 to 2 cm /yr on average) due to the convection currents in the interior of the earth. The plate boundaries move away from each other (divergent boundaries) at some places while they converge and collide against each other (convergent boundaries) at other places. In regions, they slide past each other along fault planes (transform faults). The plates are about 80 km thick. The idea of plate movements has first been proposed in 1960 and is known as plate tectonics. Evidences from several aspects collected on different ocean floors strongly confirm the plate movements and have provided enough data regarding the rate and directional movement of these plates.

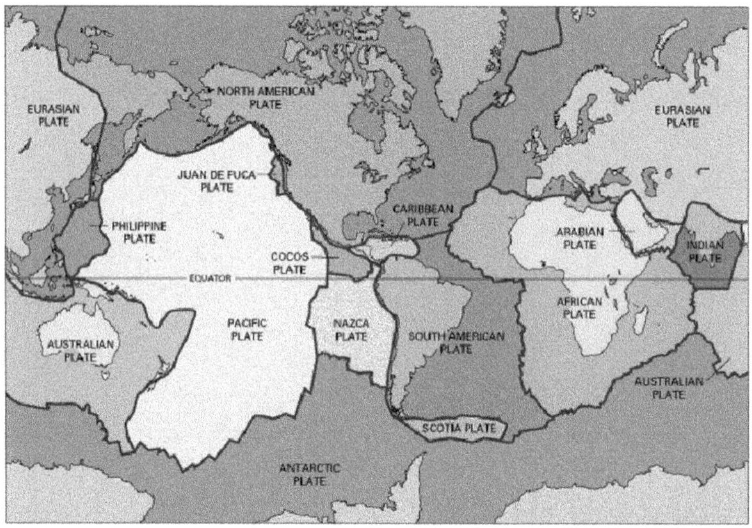

Fig. 10.1 Tectonic plates of Earth's crust. (Courtesy : U.S.Geological Survey).

At the divergent plate boundaries, new crustal material comes up as the plates move away from each other (Fig. 10.2). At the convergent boundaries, the plates collide against each other with one of them going underneath the other. This zone is known as a subduction zone. The crustal material that goes down gets molten and the magma comes up through the cracks as lava resulting in the formation of volcanic islands. This crustal disturbance also generates deep-seated earthquakes with their foci at large depths.

It is very significant to note that a large cluster of earthquake epicentres and centres of volcanic activity of all the past occurrences are confined to the plate boundaries.

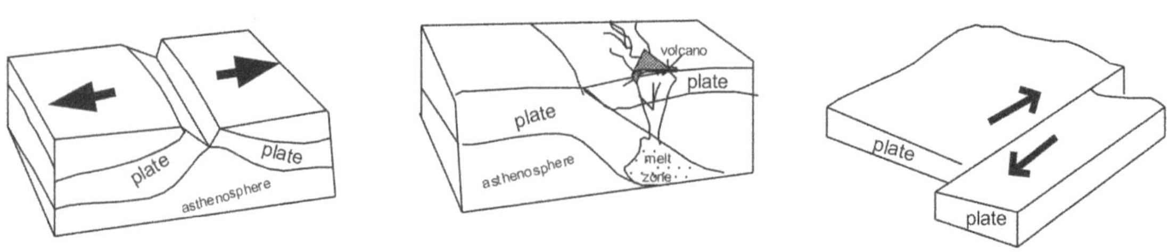

Divergent **Convergent** **Transform**

Fig. 10.2 Tectonic plate boundaries.

Earthquakes

An earthquake is a sudden motion of the ground. In the Earth's crust, at different places stresses accumulate causing slow and continuous deformation (elastic deformation) of rocks. As the stresses exceed the elastic limit of the rocks, the rocks break because of their brittle nature. This break is a fault plane along which the displacement takes place. This theory known as the elastic rebound theory, proposed in 1906, explains the origin of earthquakes. The rock behaviour involved during the accumulation of stresses is illustrated in Fig 10.3. During this sudden breakage, the strain energy is released in the form of shock waves. The focus is the place where this slippage has initially taken place. Vertically above the focus, the location on the ground surface close to the focus is known as epicenter of the earthquake.

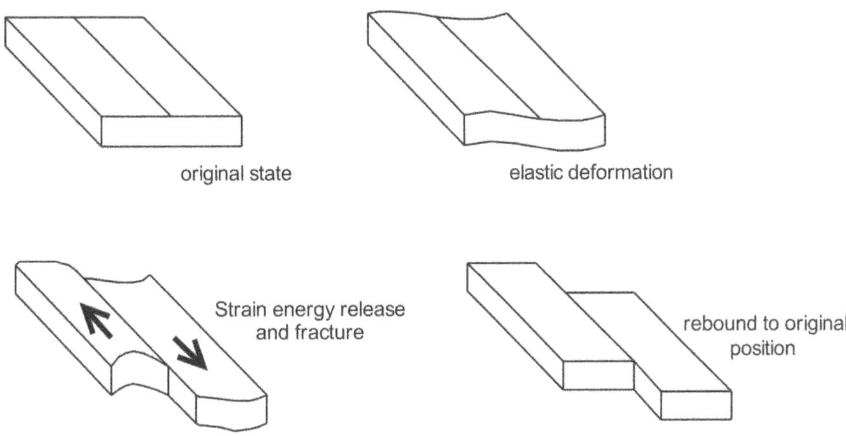

Fig. 10.3 Steps in elastic rebound process.

Recording of the shock waves is done with the help of a seismograph. The record (chart) is known as a seismogram. Different observatories located at several places in the World record the arrival of the seismic waves generated by an earthquake that occurs in any part of the globe. These records for an earthquake are useful in locating the epicentre of the earthquake and also to define the size of the earthquake.

In civil engineering practice, the response of the structures (buildings, dams, bridges, transmission towers etc) due to the ground motion is a concern for their stability during an earthquake. The response of a structure is recorded through strong-motion recorders conveniently placed at different locations of a structure so as to provide adequate information. The strong motion spectra are very useful in understanding the response of the structure due to ground motions in an earthquake. On the basis of the knowledge gained from these strong-motion records from a variety of structures, earthquake-resistant designs have been evolved. Since earthquakes can be neither predicted nor prevented, the only way to live with them is to construct structures resistant to earthquakes. It is now a common practice to use building codes incorporating this aspect. In India, it was earlier believed that only the Himalayan terrain is earthquake-prone and that a substantial portion of the Peninsular shield is aseismic. But after the Koyna earthquake of 1967 and other later shocks in the Peninsular part of the country, it is now clear that no region in India is aseismic and all the structures need earthquake-resistant designs.

Depending on the depth of focus, earthquakes are classified as

- shallow (up to 70 km)
- intermediate (70 – 300 km)
- deep (300 – 700 km)

About 90% of all the past quakes belong to the first category. As already indicated, the earthquakes of the second and third categories are related to subduction zones. At depths beyond 700 km, the material being plastic in its behaviour, earthquakes are not expected to occur.

Locating the Epicentre of an Earthquake

Epicenter of an earthquake can be located on the basis of seismic records (seismograms) obtained from three different observatories. From the location of each of these observatories on the map, arcs are drawn with the respective time-lags of shock arrival times ($t_P - t_S$) as the radii (Fig. 10.4). These arcs intersect at one point, which marks the epicenter of that earthquake. As is obvious, it is necessary to select three observatory locations divergent in direction from each other to get the spread of arcs from different directions.

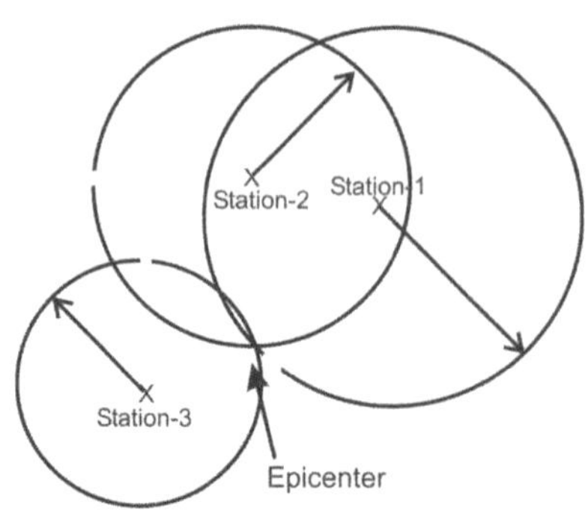

Fig. 10.4 Location of epicentre.

Size of Earthquakes

The size of an earthquake is defined by its intensity and magnitude. Intensity is expressed based on the degree of destruction caused and hence varies from place to place. It is maximum around the epicentral area. At each place, the intensity is determined based on the Modified Mercalli Scale (Table 10.1).

Table 10.1 Modified Mercalli Scale of earthquake intensities

Intensity	Effects
I	Detected only by very sensitive instruments
II	Felt by few persons at rest, particularly on upper floors of buildings
III	Felt by persons indoors, standing automobiles disturbed slightly, vibration similar to a moving truck
IV	Felt by many persons indoors, also by few outdoors; windows and doors rattle ; loose objects disturbed
V	Felt by almost everyone; bells ring; breakage of dishes; pendulum stops ; wall plaster breaks
VI	Felt by all; falling of plaster; walls crack; heavy furniture disturbed
VII	Persons run outdoors; tremors felt by persons in moving Vehicles, slight to moderate damage in well-built structures
VIII	Considerable damage to ordinary buildings; falling of walls; ejection of sand and mud in small quantities ; changes in well water
IX	Ground cracks, breakage of underground pipes; considerable damage to buildings; buildings shifted off foundations
X	Bending of rails; most of the structures damaged, occurrence of Landslides, ground badly cracked
XI	Buildings destroyed; broad open ground fissures develop, rails badly twisted
XII	Total destruction; surface displacements; objects thrown into air

Once the intensity values are assigned to different locations of the region on a map, equal intensity contours are drawn. These are known as isoseismals. The intensity at a place depends on several factors such as distance from the epicentre, depth of focus, geological formations and also on the type of construction of a structure. Since the intensity value is fixed on the basis of collection of data in the field from several sources including experiences of persons, quite often it is more subjective.

On the contrary, magnitude of an earthquake does not vary from place to place. Magnitude is a function of the energy released in an earthquake and is commonly expressed as Richter's magnitude. Relation between magnitude (M) and energy (E) is expressed as

$$\log_{10} E = 4.4 + 2.14\, M - 0.054\, M^2$$

E is obtained from the expression :

$$\sqrt{E} = c(a/h).\ (d^2 + h^2)$$

E = total energy released (ergs)

c = constant (taken as 0.625)

a = ground acceleration

d = distance (km) of the recording station from the epicentre

h = depth of focus (km)

Richter's scale has magnitude numbers up to 10. But the maximum known magnitude is around 9.6 only. Beyond this value, the earthquake energy is so high that there is total destruction. To provide an idea of the energy released in an earthquake, a quake of 6.0M involves energy of around 2.5×10^{20} ergs (equal to that of an ordinary atom bomb) while for 7.0M earthquakes, it is around 80.0×10^{20} ergs (equivalent to that of a H-bomb). For an earthquake of 8.0M, it is around 2500×10^{20} ergs. The energy increases by a factor of 30 for each increment in the magnitude. In Richter's scale, the amplitude increases on a logarithmic scale with the increase in magnitude. The magnitudes of some of the important earthquakes in India are given below :

Shillong	(June 12, 1897)	8.7
Kangra	(April 4, 1905)	8.0
N. Bihar	(Jan 15, 1934)	8.3
Assam	(Aug 15,1950)	8.6
Koyna	(Dec 11, 1967)	6.4
Uttarakashi	(Oct 20, 1991)	6.5
Killari (Latur)	(Sept 30, 1993)	6.5
Jabalpur	(May 22, 1997)	6.0
Gujarat	(Jan 26, 2001)	7.5
Andamans	(Dec 26, 2004)	9.0

Since 1990, 12 earthquakes with magnitude of 8.0 and above have been reported in different parts of the World. Details are as follows :

1994 (June, 9)	N.Bolivia	8.2
1994 (October, 4)	Kuril Islands	8.3
1995 (July, 30)	N.Chile coast	8.0
1995 (October, 9)	Jalisco coast, Mexico	8.0
1996 (February, 17)	Irian Jaya, Indonesia	8.2
1998 (March, 25)	Balleny Islands	8.1
2000 (November, 16)	New Islands Region, P.N.G.	8.0
2001 (June, 23)	Peru Coast	8.4
2003 (September, 25)	Hokkaido, Japan	8.3
2004 (December, 23)	N. of MacQuairie Island	8.1
2004 (December, 26)	Off W.Coast, N.Sumatra	9.0

Magnitude is also expressed as a moment magnitude. While Richter's magnitude is based on the energy, moment magnitude is closely related to the cause of the earthquake than to only the energy. It is expressed as :

M_m (moment magnitude) = $2/3 \{\log_{10} M_0 - 16\}$

M_0 (dyne-cm) = (strength of the rock) × (fault area) × {total displacement along the fault plane}

Strength of the rock is in terms of elastic modulus. The slip along the fault plane is inferred from the direction of the first arrival of P wave from the seismic record (its direction whether up or down and its amplitude). Once several records from different adjacent stations are interpreted in this manner, the fault plane orientation and the slip can be estimated. This is the procedure commonly adopted, since the fieldwork for fault details at the epicentre will not be of any help as the destruction in the area would be considerable.

Seismic Zoning Map of India

On a seismic map, the country is divided jnto several zones in terms of severity of expected ground motion. The zoning map is reviewed and revised periodically based on additional data available. A brief account of significant features of the seismic zoning maps in the past has been presented by Jai Krishna (1992).

The first Zoning Map (1962) was prepared on the basis of historical data available regarding the occurrence of earthquakes all over the country and their severity (magnitude). This was based primarily on the epicentral map and on the data of isoseismal maps published by the Geological Survey of India for some strong earthquakes and the earliest zoning map brought out by G.S.I. in 1935. Epicentres of earthquakes with magnitude 5 and above were plotted. Six zones were indicated. Zone I is least severe and the zone VI is the most severe. The entire Peninsular India except for few areas was shown as zone-0 and considered aseismic (very stable region free from earthquakes). Subsequently the zoning map was revised in 1966. Zone-0 was narrowed in its extent. Koyna earthquake in 1967 has confirmed that Peninsular India is no longer asiesmic. In the revised map of 1970, zone-0 was accordingly removed. Many of the areas in zones V and VI were merged into one zone because of their high risk. So the number of zones got reduced from seven to five. In the zone map brought out in the year 2000, the earlier zones II and I were merged. Additional regions in South have been included in zone III. Zone IV includes areas of high risk. Entire NE region, parts of Uttaranchal, Rann of Kutch (Gujarat) and Srinagar are included in zone V.

Earthquake Hazards

1. ***Destruction of buildings due to severe earth-shaking (ground motion) :*** This is common in several earthquakes. Vertical and partial collapse of a building during the Bhuj earthquake (7.5 M) on January 26, 2001 is indicated in Fig. 10.5.

Fig. 10.5 Vertical split and partial collapse of a 5-storey reinforced concrete frame building in Bhuj during earthquake on January 26,2001.
(Courtesy : EERI Photo : Challa V.R.Murty I.I.T.,Kanpur)

2. ***Ground fissures and wide openings*** : Often sand and mud with water is ejected. For example in Chamoli earthquake (Uttaranchal) in 1991, fissure patterns were in two prominent directions :
   (i) Along NNW-SSE (as in the vicinity of Makku-Hafla region, with the fractures extending up to about 500 m in length with 40-50 cm width and 10 m of depth. A movement of the block on SW up to 20-30 cm downwards was noticed.
   (ii) E-W trending fractures as around Mayali area, extending to a length of 60 m. In this case, the downward displacement of the south block was observed.

   During the earthquake on December 26(8.9 M), a part of the Aceh coast has been ripped apart (Fig.10.6).

Fig. 10.6 Ripping apart a part of the northeastern coast of Aceh, Indonesia. (Earthquake of Dec.26, 2004) Photo: AP

3. ***Dislocation of transportation routes (highways, bridges, railway tracks etc) :*** During the California earthquake of 1994, a two-level freeway has been damaged. In the Uttarkhasi region, the metal structure of a road bridge has been very violently twisted and damaged.

   During the disastrous earthquake (8.7 M) in Alaska on March 27 of 1964, lateral displacement of the deck of a road bridge (Fig. 10.7) has taken place.

Fig. 10.7 Lateral displacement of Copper River Bridge during Alaska earthquake of 1964 (Courtesy : U.S. Geological Survey Photo : George Plafker)

A vertical displacement of a roadway (Fig. 10.8) has resulted. In the figure, the pavement in the foreground has gone down up to around 1 m.

Fig. 10.8 Vertical displacement of a road SE of Ingram Creek, Alaska during the 1964 eathquake Photo : Helen Foster. (Courtesy : U.S. Geological Survey)

Fig. 10.9 Damage of abutment of a 2-lane highway Bridge near Bhachau during the Bhuj earthquake (January 26, 2001) (Courtesy : EERI Photo: Sudhir Jain I.I.T., Kanpur).

4. *Liquefaction :* In presence of water, soils under dynamic state of vibration behave like fluids. This behaviour is termed as liquefaction. Settlement of buildings and other structures is common due to this (Fig 10.10). In some instances, the water through the soil is forced out due to cracking of the surface and its oscillation. This ejection of water with sand forms sand boils. Typical sand boil developed during Bhuj earthquake of 2001 is illustrated in Fig. 10.11.

Fig. 10.10 Subsidence and tilting of buildings due to liquefaction during the earthquake in Nigata (Japan) in June ,1964.
(Courtesy : U.S.Geological Survey)

Fig. 10.11 Sand boil due to liquefaction during Bhuj earthquake (Jan. 26, 2001).
(Courtesy : EERI Photo : Bob Wright, Mill Valley)

Structural damage during liquefaction is due to the following causes :

A. Lateral separation of ground
B. Increase in the lateral stress in liquefied soils
C. Reduction of bearing capacity of soils
D. Formation of cavities (sink holes) due to sand boils
E. Development of high pore pressure resulting in failure of soils

5. ***Quick clays :*** Quick clays, present with excess of pore water, are responsible for large subsidence and landslides. Typical example is the Alaska (Anchorage) earthquake in 1964.

6. ***Landslides :*** Mass movement along the unstable hill slopes takes place resulting in massive slides on the hill slopes. Heavy debris sliding down in large quantity into a river valley creates blocking the flow and flooding of the valley. This results in heavy loss of life and property in the region. Slumping of mass due to slope failure is illustrated in Fig. 10.12.

Fig. 10.12 Slumping due to slope failure during Bhuj earthquake of January 26, 2001. (Courtesy : EERI Photo : William Lettis-William Lettis and Associates)

7. ***Tsunamis and seiches :***

Tsunamis are generated by submarine earthquakes, which are common in subduction zones. In the subduction zones off the coasts of Mexico, Nicaragua, Chile and Indonesia, around 17 tsunamis have been reported in a 5-year period (1992 to 1996). The word "tsunami" is originally from Japanese and it literally means a harbour wave. These high-energy waves are caused by rapid displacement on the sea floor during the earthquake. These are large waves with enormous wavelengths and travel several thousands of kilometers with very little loss of energy due to their high momentum. They are much more damaging compared to the ordinary wind-generated tidal waves. While tidal waves are confined to near-shore shallow depths, tsunamis extend across the oceans. They have low crestal heights in deep-water zones and only while approaching the land, they build up with large amplitudes.

Tsunamis assume different forms depending upon the geometry of the sea floor. They can be sudden and rise to great heights while reaching the coasts. In some instances, the sea withdraws in the first stage. Immediately the water tosses back in the form of large tsunami. For example, exposure of new structures related to the Mamallapuram (Mahabalipuram) temple complex in the region of retreat of the sea just before the tsunami hit were reported. This was followed by the large tsunami waves hitting the coast resulting in the submergence of the exposed structures there, once again under the sea.

For the first time in 2004, tsunamis have caused disaster across the coasts of Thailand, Myanmar, Andaman and Nicobar Islands, East and West coasts of India and even as far as Somalia and Kenya. These waves have been generated due to an earthquake of magnitude 9.0 which occurred on December 26, 2004 off the West Coast of N. Sumatra in Indonesia.

The epicenter was located at 3.298^0 N and 95.779^0 E of latitude and longitude respectively and the focal depth was reported as 10km at the interface between the Burmese and the Indian Plates. The Indian Plate is reported to be moving at 5cm/yr in a NE direction relative to the Burmese plate. This earthquake has resulted in a N-S fault of approximately 1000km length with a slippage of about 15m at the subduction zone with the Indian Plate underneath the Burmese Plate. It is interesting to note that the tsunamis were generated in the E-W direction, transverse to the fault line. The maximum destruction has been in Sri Lanka and India (Andaman-Nicobar islands and several areas on the main land East Coast which include Cuddalore, Nagapattinam, Velangini and Chennai in Tamil Nadu as also areas of Nellore , Prakasam , Krishna and East Godavari Districts of Andhra Pradesh).

Although not on the direct line of incidence of the tsunami, parts of Kerala coast have also been affected by the wave, due to the diffraction effect.

Seiches are also generated during earthquakes. These are generated in standing or restricted water bodies. These are low-amplitude waves developed during the oscillation of water. The fundamental period is a function of the magnitude of causative earthquake and the depth of the water body.

8. ***Fires :*** Fires are common due to the breakdown of power lines
9. ***Breakdown of sewer lines*** causing epidemic in the quake-affected areas due to pollution and unsanitary conditions.

Stability of Structures during Earthquakes

Amount of damage suffered by a structure is dependent on several factors such as the ground motion, duration of the earthquake, its magnitude, site conditions, type of materials used in its construction and the design of the building.

Horizontal ground motions play a significant role in the stability or damage of any building during earthquakes. They result in lateral accelerations, which impart inertial forces on the structure. The stability of any structure can be assessed by considering the peak lateral force. In low-rise rigid structures, the peak lateral force is the product of its mass (M) and peak ground acceleration (a_g). For flexible structures, this lateral force would be less than the product ($M.a_g$) as some energy is absorbed in its bending. In tall and very flexible structures, the lateral force even exceeds $M.a_g$ due to the increase of inertial forces.

During an earthquake, both the structure and the foundation material vibrate in response to ground motion. The time taken to complete one cycle of oscillation is known as the fundamental period. The fundamental period for the structure (T_{str}) and the fundamental period for the soil (T_{soil}) are both involved in their response to the ground motion. For buildings, it varies with the type of structure from low to high values for low to tall buildings. The fundamental periods for the foundation strata vary, ranging from about 1sec for hard rock to several seconds for loose unconsolidated and deep soils. Once both these periods for the structure and foundation material are similar, response amplification occurs which leads to violent shaking and ultimate destruction of the structure. Thus tall buildings located on soft soils suffer intense damage.

Duration of an earthquake is important. For example, short duration motion with high amplitudes may cause less damage compared to a long duration motion even with moderate amplitudes. This is due the degradation of stiffness of the structure caused by more load reversals in the long duration motion. The duration of a quake generally increases with its magnitude.

Type of construction influences the damage. In general, small wooden frame structures well anchored to the foundations resist earthquake shaking very well. Well-nailed and jointed wooden frames, being more stiffened, withstand ground motion with less damage. The safety of RCC and steel frames is critically dependent on the design detailing and construction. Non-reinforced brick and mortar constructions are most vulnerable for earthquake shaking (Fig.10.13). Due to the weakness of mortar, cracks first appear in the same, ultimately resulting in the wall failure. Buildings with a distribution of weight at the bottom are more suitable compared to the top heavy ones. In the latter, higher overturning effects are developed due to the swinging of mass at higher elevations. Cross-bracing with steel in tall high-rise buildings enables the building to withstand lateral motion due to the stiffness in the structure. The use of strong RC structural walls also provides greater resistance to earthquake shaking. Thus, there are four virtues in earthquake-resistant structures, namely at least a minimum lateral strength, sufficient stiffness, large ductility and good seismic structural configuration. These are specified in the Indian structural codes. Deficiency in any of these can result in poor earthquake performance (Fig. 10.14, 10.15, 10.16). Many existing structures designed as per earlier Indian seismic codes may require retrofitting to become earthquake-resistant.

Fig. 10.13 Collapse of a house with stone masonry and mud mortar at Anjar during the Bhuj earthquake (January 26,2001). (Courtesy : EERI Photo : Sudhir Jain I.I.T., Kanpur)

Fig. 10.14 Buckling of longitudinal steel due to opening and snapping of lateral ties during the Bhuj earthquake of January 26,2001. (Courtesy : EERI Photo : Challa V.R.Murty I.I.T., Kanpur)

Fig. 10.15 Corner column failure in Karpos in Skopje, Yugoslavia during the 1963 earthquake (6.5M) (Photo: Glen V. Berg)

Fig.10.16 Detail of column failure

In Japan, through continuous analysis of the response spectra of various structures, the quake-resistant codes are evolved enabling the structures to withstand during strong ground motion. In chemical and other industrial establishments, the dangers are averted by the immediate computer-operated controls either through safeguarding the flow of chemicals or by opening alternate and safe channels for flow if the original installations are damaged.

In waterfront structures (dams), in addition to the inertial effects due to their own mass, hydrodynamic effects due to the adjoining water (either static or flowing) should also be considered.

Seismic Risk Zone Maps

Seismic risk maps are prepared for different areas by plotting of all quake related data. Such data include the type of geological material, active faults, concealed faults, soils with liquefaction potential, potential areas for landslides etc. Details regarding the epicentres of past earthquakes, if any, are also incorporated on the same. For preparation of such a risk zone map, GIS is useful as a tool to integrate all the thematic maps and to rank the area for hazard zoning.

Vulnerability data also can be incorporated in such maps to indicate highly potential and critical areas such as hospitals, military establishments, public utility systems, and transport networks.

The seismic risk maps help in locating future infrastructure.

Volcanic Hazards

Volcanic hazards encompass potentially dangerous volcanic processes. As already indicated, volcanoes and earthquakes are associated with tectonic disturbances at the plate margins. The Earth's crust is highly fractured at these margins. These fractures extending to depths facilitate migration of magma upwards. The rate of travel of magma depends on its composition. Magmas vary in their composition from those rich in silica (granitic magmas) to those deficient in silica and richer in Fe and Mg constituents. Silica-rich magmas are very viscous and their movement is relatively slow. This explains the rare occurrence of volcanoes with silica-rich lava. On the other hand, basaltic magmas travel faster (being less viscous) and if erupted as volcanoes, the lava flow spreads to considerable aerial extent. Deccan trap basaltic flows, covering extensively the State of Maharashtra and the border regions of Andhra Pradesh, Madhya Pradesh and Karnataka, are classic examples of the same.

The molten rock material (magma) emerges on to the surface as lava. Volcanic activity involves eruptions with ejection of lava along with several volcanic gases. The nature of eruption depends on the viscosity of the molten material and the quantum of gases in it. At high pressures underneath the earth's surface, the gases are dissolved in the magma. As the magma travels up, these gases gradually get released from the melt due to decrease in the pressure. This results in volume changes (because of reduction in density). This mechanism explains the movement of the molten material and the violence of eruption. The volcanoes with low-viscosity material erupt with less violence. Because of the spread of the material, the top of the volcano is relatively broad. Such volcanoes are termed as shield volcanoes. The eruptions with high viscosity material give rise to large volcanoes with steeper sides. Such volcanoes are silent for longer periods between two successive eruptions. These volcanoes are known as strata volcanoes. With very violent eruption, if the top is blown away, a collapsed structure, termed cauldera, is formed. Subsequently water may fill the cauldera resulting in a crater lake. A crater lake formed through such process in Oregon state (U.S.A.) is shown in Fig. 10.17.

Fig. 10.17 Crater lake, Oregon , U.S.A. (Photo: Gokhale, 2004).

The only active volcanic spot in India is situated in the Barren Islands, 130 km NE of Port Blair (S.Andaman Islands) in Bay of Bengal. (Fig.10.18). This is located in the seismotectonically active belt on the converging plate margin of Indian Ocean.

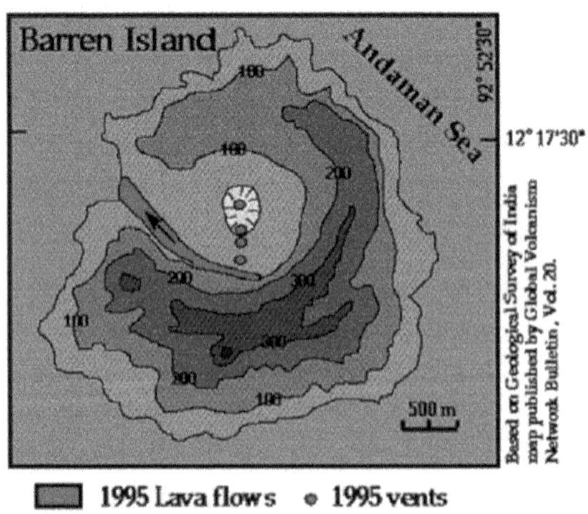

Fig. 10.18 Barren Islands, Bay of Bengal.
(Courtesy : Global Volcanic Network, Smithsonian Institute,U.S.A.)

Up to 1900, seven eruptions have been recorded. Activity was noticed in March 1991, which continued till November 1991. This eruption was monitored with the help of temporal satellite data. Vent temperature was estimated to be around 1200 ^{0} C. The somma has been reported to be around 1.6km in width. Subsequent eruptions were in 1991 and 1994, which lasted from December 1994 to June 1995. Initial activity, which started from a vent on upper NE flank of the central cone, was later reported to have expanded to the main crater.

Types of Volcanic Hazards

Volcanic eruptions are hazardous and occur in many forms. The details are as follows :

1. ***Volcanic gases :*** When a volcano erupts, dense cloud of gases spreads into the stratosphere. Prominent among them are water vapour , carbon dioxide (CO_2), sulphur dioxide (SO_2) and hydrogen chloride(HCl) with subordinate amounts of hydrogen flouride(HF),carbon monoxide (CO) and hydrogen sulphide (H_2S) .

   Sulphur dioxide contributes to the acid rain and also causes depletion of the ozone layer. Carbon dioxide, being heavy, concentrates in the lower reaches of the atmosphere. This is a lethal gas and is a threat to population. Accumulation of CO_2 on the bottom of tropical lakes in massive amounts, causes enormous damage once such a lake empties. It has been reported that such a process in Lake Nyos has been responsible for 1500 deaths and injuries to tens of thousands of persons. Flouride and chloride gases contaminate water. They cause skin irritation.

2. ***Lava flows :*** Lava flows, being hot, are very disastrous. They destroy everything on their way. Huge piles of lava bury the property beyond repair and reclamation. This is the basic difference compared to landslide hazard damage to structures which can be repaired. Quite often, methane gas is generated from the buried vegetation. This gas is liable for explosion on catching fire.

   Volcanic flows vary in their temperature and size. The temperatures vary between 200^0 C and 1000^0C, causing extensive burning of all the material they encounter. They travel with velocities ranging from 30 m/s to 200 m/s and can extend as far as 200 km. The flows follow stream valleys and sometimes block the streams, resulting in an overflow in the subsequent flood season. The flows are also responsible for floods from the melt water in a terrain with snow or ice.

3. ***Volcanic fragments :*** Also called tephra, the material varies in size. Volcanic ash contains the smallest size fragments (< 2 mm). The ash spreads as a cloud covering enormous areas following the wind direction. Particles smaller than .01 mm have been reported to stay in the atmosphere for as long as 2 years. This cloud reduces the visibility and is slippery with water. In a direct blast, right above the volcanic vent, the thrust zone grades into a convective zone wherein the prevailing heat enables the direct lifting of the material up into the stratosphere. The material there gets dispersed by the strong winds. This material sometimes gets electrically charged. Lightning produced due to such charges is believed to have been responsible for the death of several persons. Volcanic ash causes breathing problems. Ash deposited in considerable amounts on the top of flat roofs cause the collapse of buildings.

   The larger fragments of sand and gravel size are known as lapilli. Fragments of larger lapilli are called blocks. Very large blocks are known as bombs. They all contribute to the violence involved in an eruption. All these settle down closer to the volcano in its vicinity.

4. ***Lahars :*** Volcanic material mixed with water forms a slurry of consistency similar to wet concrete mix. The water in this mix is contributed by heavy rainfall or from melting of snow. Lahars containing around 80% of the volcanic material are known to have smooth laminar flow with velocities as high as 40m/s. They are very disastrous as they carry heavy load of different dimensions including big uprooted trees. They destroy bridges and buildings.

Lahars flowing in the stream valleys often undercut the valley slopes. Due to the resulting slope instability, structures on the top of the slopes are destroyed. For example, lahars developed on eruption of Mt. Rainier in the Washington State (USA) in 1990, St. Helens in 1982 (Fig. 10.19), Casita volcano in Nicargua in 1998 and Mt. Pinatubo in 1991 as also in 1994 have caused extensive damage.

Fig. 10.19 Lahar (dark area in the middle) flowing on St. Helens Washington State,U.S.A., March 1982. (Photo: J.Casadwall Courtesy: U.S.Geological Survey)

5. ***Volcanic landslides :*** Also called debris avalanches, these involve movement of debris of different sizes down the slope. The slides are generated by sudden blowing out of a side of the volcano. Such slides, of large dimension, have high momentum and move very fast.

6. ***Volcanic earthquakes :*** These are due to stress changes resulting in the rocks due to the movement of the magma. Quite often, volcanic tremors are experienced prior to an eruption. The volcanic earthquakes cause considerable damage to the buildings and structures.

Volcanic Hazard Map

Based on the survey and collection of data in a region around the volcano covering enough distance, the effects of the volcano in terms of the extent of damage caused due to the lava flow, fragments, slides and lahars can be represented on a map. On such a map, delineation of the zones of varying damage is done in terms of zone 1 (with highest damage) to zone- 4 (lowest damage). This helps in visualizing the type and extent of potential damage that can be inflicted by the volcano in a subsequent eruption.

Monitoring of a volcano from time to time would enable understanding of the nature of changes preceding an eruption.

Mass Movements

Movement of geological material takes place in different forms. Under the influence of gravity, movement of rock and soil materials takes place down the slope. The type of movement depends on the composition of the material, degree of presence of water and the steepness of the ground. The movement can be either extremely slow (creep) or it can be very rapid (mud flow or a debris avalanche).

Movement of material can also be vertically down. This type is known as subsidence. Sinking of the ground takes place due to underground open space being present. This may be due to natural causes or induced by human activity. Subsidence can be slow or relatively fast depending upon the type.

Knowledge of the areas susceptible for mass movements is essential in planning of civil engineering projects. Delineation of potentially hazardous slopes or areas of subsidence is a vital part in the planning stages. Mapping of hazardous areas involves data from several aspects including the field mapping. Databases incorporating various operative parameters are integrated to arrive at a final map indicating zones of differing susceptibility to mass movements.

Classification of Mass Movements

Type of Movement	Nature of Material	Time-span involved
Falls		
Rock Falls	Rolling of rocks along with soil	Sudden
Slides		
Tabular	Rock slabs sliding along joint surfaces and bedding planes	Sudden
Rotational	Sliding of unconsolidated material along slip surfaces – due to heavy rains or earthquakes	Fast-sudden
Flows		
Dry	Movement of sand dunes (in arid regions)	Several km /hr
Wet	Creep Slow imperceptible movement	Up to 1cm /yr
Earth Flows	Movement of earth material due to heavy rains	Up to 100 m /day
Debris Avalanches	Sudden collapse of large mass of earth material (caused by volcanic activity or earthquakes)	Sudden
Slurry Flows		
Debris Flows	Flow of debris caused by heavy rains	Up to 100m /hr
Mud Flows	Flow of slurry of debris, due to heavy rains. If the flow is generated due to volcanic activity , it is called a lahar	Several km /hr
Solifluction	Soil saturated with long-standing water slipping along a frozen subsurface	Few cm /yr
Subsidence		
Natural	Collapse of surface due to sink holes (solution cavities) in limestone area	Slow-sudden
	Oxidation of organic soil material consequent to groundwater fluctuations in marshy areas	Slow
Man-made	Collapse of surface due to underground cavities without adequate supports in coal mining.	Slow-sudden
	Ground subsidence due to extensive underground fire and burning of coal	Slow
	Subsidence caused by excessive ground water pumping (hydro-compaction)	Slow
	Ground subsidence in oil fields due to removal of oil and gas	Slow

Types of Mass Movement

Falls

These involve sudden and rapid rolling and falling of the loosened mass consisting of rock fragments and soil. As falls are sudden and the size of the rock fragments quite often is in the boulder range, falls are often disastrous.

Flows

As shown in the above table, the nature of a flow varies with water content and consistency of the material.

Creep

It is a slow movement of the material down the slope. The movement is so slow that it is imperceptible. Individual particles move independently and hence the slope movement is not as one unified material. Evidences of creep are often recognizable in the field in terms of the shape of vegetation or any other installation (like electric or telephone poles) located on the slope. Since the time scale involved is so large (with very slow movement), the tree trunks exhibit a gradual curved shape. This is because of the tendency for the trunk to remain vertical on ever-sliding slope. Similarly, the tilt can be evidenced in the location of utility poles.

Solifluction

The rate of movement is relatively faster than creep. This involves flow of water-saturated material along an underlying frozen material (permafrost). When the top material of a frozen mass starts melting in summer, the resulting melt water cannot infilter to the frozen subsurface (which is impervious). This waterlogging in the top layers results in slippery situation resulting in the downward motion of the upper layers. This is common in temperate climates.

Debris flows, earthflows and mudflows

These are caused due to heavy rains. The differentiating characters in terms of time-span of movement and the type of material involved are indicated in the above Table. A debris flow involves soil and rock mixture with a substantial fraction having grain size coarser than sand. Mudflows and earthflows have relatively restricted range of sizes with sand and clastic particles. Mudflows have greater water and the slurry is quite thick in its consistency. Mudflows can also be created during a volcanic eruption. Heavy rainfall is quite common during the eruption period. The volcanic material combined with considerable water forms thick slurry, which flows into the valley and follows it. These flows are called lahars and are quite disastrous as they pick up load of considerable sizes like big uprooted trees, which are also carried along.

Grain flows

Common in arid regions, these are involved in migration of sand dunes. The particle sizes are in the silt-sand range and the mass is relatively dry.

Debris avalanche

Sudden collapse of a part of a mountain or mass of a huge dimension, commonly due to earthquakes or volcanic eruptions. As they are sudden, the resulting damage is considerable.

Slides

Rock slides : These involve movement of tabular blocks of rock, loosened by erosion along weak zones. Sliding takes place along the bedding planes or joints. These are sudden and quite often involve individual rock blocks of enormous dimensions.

Land slides and slumps : The sliding takes place along a slip surface. This surface is curved with the depth decreasing towards the bottom of the slope. In plan, the slip surface is arcuate. The material is largely of soil with rock fragments. On a hill slope with vegetation, the slide area is quite conspicuous from stripping of vegetation. If the slide is not a recent one, still the tonal contrast in the vegetation is discernable. Landslides are caused by heavy rains, blasting vibrations as also earthquakes.

Causes of landslides : Failure of slopes is contributed by several factors:

- *Water content :* If there is no adequate drainage, the water saturation contributes to reduction in intergranular friction, increase in the weight of the earth material and also to the lubrication resulting in the slipping of the material down the slope. Water is the most important contributing factor. Pore pressure increases in the material due to ineffective drainage.
- *Slope :* Stable slope angles vary with the type of material involved. This angle is known as angle of repose. For coarse materials, this angle is around 30^0.
- *Nature of the material :* The lithology of the material influences a slide. Internal friction and cohesion contribute to the opposing force to the movement. As evident, highly weathered and fractured materials are susceptible.
- *Geological structure :* Bedding planes or joint planes dipping towards the slope add to the instability of the slope.
- Earthquakes
- Blasting vibrations; quarrying.
- Loading on the top due to constructions
- Undercutting at the toe of the slope: Quite common in the river channels.

Slope Failure Mechanism

Forces acting along the slope can be divided into two categories, namely the driving forces and the restoring forces. The former contribute to the downward movement while the latter oppose and act against the direction of the movement.

The weight of the material (W), acting vertically downwards, can be resolved into two components : W Sin θ acting along the slope down and W Cos θ acting at right angles to the slope.

The driving force (D) = $W \sin\theta$

The resisting or opposing force (R) = $(W \cos\theta - p) \tan\theta + C$.

C = cohesion, p = pore pressure, and θ is angle of shearing resistance

The ratio R/D is the factor of safety. For a slope to be stable, this ratio should be greater than 1.0.

Prevention of Landslides

Following measures help in preventing landslides:

1. *Provision of adequate surface and subsurface drainage :* Providing horizontal and vertical drains as also drains at the toe. The intent is to enable water to freely drain out of the material.
2. Construction of retaining walls
3. Driving cylindrical piles or providing pier walls
4. Modifying the slopes to stable angles
5. Growing vegetation to hold the material together.

Land Slides in India

Landslides are reported in the hilly terrains in different regions of India .The most disastrous slides that have taken place in recent past are in the Himalayan terrain in the North and the Nilgiri hill region in the South.

Landslides in the Himalayan Terrain

The landslides in the Himalayan region are mostly confined to the South boundary of the Tertiary folding belt (Choubey, 1998). Overthrusting and fracturing are quite common in this belt. The region is an active seismic belt.

Disastrous landslides have occurred in the Garhwal region of Uttaranchal. The Alaknanda and Bhagirathi valleys are prone to slope instability. The Alaknanda region is geologically composed of a sequence with the lowest group (Dudhatoli group) consisting of quartzites and phyllites separated from the overlying quartzites, schists, slates and carbonate formation (Garhwal group) by the North Almora Thrust. This group of formations is highly disturbed with joints and shear zones. The Central Crystalline Group overlying the Garhwal Group is composed of schists, gneisses and granites. In July 1970, heavy debris from Patalganga valley has been transported into Alaknanda. The material was so much that the flow of Alaknanda was not only constrained but the channel was shifted to the north bank. Damming of the river by debris has resulted in considerable increase in the water level of the river leading to heavy floods. On July 20, 1970 several landslides were reported to have occurred within about an hour at Helang, Belakuchi, Tangani, Pakh, Birahi and Nandaprayag. The flooding in Alaknanda due to these slides has resulted in a silt and rock fragment load of about 9 Mcum.

Another disastrous slide took place in Malpa area: This was a rock avalanche that took place on August 18,1998. Malpa village is located on the banks of Kali River (Pithorgarh District) in the Kumaon Himalayas. Huge rock fall took place suddenly. A single mass of rock was reported to have been detached and while rolling down the slope broke into hundreds of pieces. The fragments were quite big in size (up to about 4 to 5 m size). The velocity of fall was estimated to be about 25 to 30 m/s. The piled debris was around 20 m in height. The Malpa area is composed of granites in the North separated from quartzites, sericite schists and crystalline metamorphites by the Main Central Thrust trending in a near east-west direction and located around 8 km away north of Malpa. In the rock fall area, the formations are quartzites interlayered with schistose bands. The rocks are highly jointed. The rock fall was associated with a heavy dust storm. The fragments, rolling very fast during the fall, have generated sparks and bright flashes of light. (Bhandari, 1999)

In the Darjeeling and Sikkim Himalayan regions, the Tiesta river is known for massive erosion and undercutting. In 1968, in a period of three days, numerous landslides occurred (Chopra, 1977, Choubey, 1989) during heavy rainfall of about 500 to 1000 mm. The 60 km highway between Darjeeling (West Bengal) and Gangtok (Sikkim) was disrupted at more than 90 places.

In Himachal Pradesh, a disastrous slide occurred in August 1993 in Nathpa area. The slide took place along a joint plane parallel to the valley side on the right bank of Sutlej river(Choubey, 1999) leading to the obstruction of the river flow near the Nathpa project area. The slide, beside damaging a stretch of the road on National Highway, has created a lake (10 km long and 10m wide) with a reservoir impounding about 5.6 Mcum of water. It has been reported that the flow in Sutlej river was entirely blocked for about 12 hrs. The water of the lake has entered partly into the outlet of the powerhouse.

Slope failures are common in the other Himalayan terrains in Mizoram, Nagaland, Arunachal Pradesh and Meghalaya.

Recent landslides in the Himalaya terrain are listed below :

Himachal Pradesh

- Maling slide (Sept. 1968) : A bridge and stretch of a road (2 km) washed out
- Near Soldering Nala (Dec. 1982) : 3 bridges and 2 km of road destroyed
- Nathpa (Nov. 1989) : Road destroyed (about a km)
- Kullu (Sept. 1995) : Road (1 km) destroyed, 32 persons killed

Uttaranchal

- Kalisaur in Garhwal region (July 1968): Road damaged extensively
- Okhimath in Garhwal region (Aug. 1998): 69 persons killed, buildings destroyed
- Malpa in Garhwal region (Aug. 1998): 205 persons killed, road to Manasarovar damaged

Jammu & Kashmir

- Nashri (Jan. 1982) : Every year causes damage to the road
- Malori, Jammu (June 1995) : 6 persons killed, National Highway 1A damaged

West Bengal

- Kalimpong, Darjeeling (Aug. 1993): 40 persons killed, heavy loss of property

Arunachal Pradesh

- Itanagar (July 1993): 25 persons killed, road (2 km) damaged

Mizoram

- Aizwal (June 1993) : 4 persons killed, damage to property
- Aizwal (May 1995) : 25 persons killed , road extensively damaged

Nagaland

- Kohima (Aug. 1993): 200 houses and 5 km road damaged, 500 persons killed

Selected Landslide Regions in South India

Nilgiri hills, Tamil Nadu

In the Nilgiri hill region, eight areas are identified which are landslide-prone. Numerous landslides have occurred in the past in these areas. Among them, major slides are in Runnymede, Glenmore, and Coonoor areas. The slopes vary up to 30^0. In the Runnymede slide, between two horst blocks, series of grabens were reported (Seshagiri et al, 1982). In addition, many parallel and arcuate fractures were identified. Several retaining walls were constructed to keep off the debris. In the Glenmore slide area; numerous fractures have separated the area into three large slides. The causes of the landslides have been attributed to the undercutting of the hill slopes by the Katteri river and also due the water saturation of lower portions of the slopes during heavy rainfall.

Amboori landslide in Kerala

On November 9 of 2001, a disastrous slide occurred around Amboori (about 20 km from Thiruvananthapuram) located in Poochmukku in Neyyar dam on the slopes of Western Ghats. Due to heavy rains and water-logging of a high hillside with boulders, massive pile of debris came suddenly rolling down.

In Kerala, most of the slides are confined to dissected zones of plateau margins separating the plateau surfaces and the plains.

Landslide Hazard Zonation Mapping

Based on the fieldwork, geological mapping and other available collateral data, landslide hazard zonation maps are prepared for different regions. All causative factors (lithology, subsurface/ surface drainage conditions, slope, geological structural features with their relation to slope and weathered status of the material etc.) are included in the field mapping. In addition, studies on the earlier slides are also carried out. These studies involve three phases:

1. Study of the topography in and around the slide areas, type of movement, intensity of sliding, depth of the slip plane and the geometry of the slip surface.
2. Changes in the area over a period of time (using time-based aerial photographs and remote sensing data
3. Integration of all the databases to arrive at a slope instability map.

Landslide hazard zonation map is prepared by ranking the different locations all over the region for each of the parameters in the data files. The ranking system involves assigning a rank for each parameter from say 1 to 10 depending upon the contribution of the parameter to the slide occurrence. Finally, the zonation map is obtained with different zones in the region delineating the degree of susceptibility to landslides

Landslide hazard zonation atlas of India has been prepared by the Building Materials and Technology Promoton Council (BMTPC) of the Ministry of Urban Development (New Delhi) in collaboration with the Anna University (Chennai). Being GIS-based, this atlas is an extension of the vulnerability atlas (which includes earthquakes, floods and cyclones) prepared in 1996.

For the preparation of the atlas, thematic maps used are from different organizations including Geological Survey of India (for geology, neotectonic and geomorphology maps), Survey of India (for road network), National Atlas and Thematic Organization of Department of Science and Technology (for slope, landuse, rainfall, and physiography maps), Department of Ecology of Osnabruck University in Germany (for vegetation maps) and I.I.T., Roorkee (for seismic zoning maps). All these thematic data were integrated on GIS. As a first step, landslide inventory map was produced on the basis of all available information and data. In the next step, landslide hazard map was prepared by assigning weightages to different parameters. By repeated comparision of both these maps and continuous adjustments of the weightages, the best matching hazard map was obtained. The hazard zonation maps are very useful in the planning for disaster mitigation. They can be used as base maps in large-scale hazard mapping of landslides

Subsidence

Subsidence involves vertical collapse of ground. It can be either a slow process or a sudden feature. Subsidence results in depression on the ground surface.

Subsidence can be caused by natural process or through human activities. Natural processes responsible for subsidence

1. ***Carbonate dissolution in the subsurface***

   This is common to limestone terrains. If the ground water is slightly acidic, it reacts with $CaCO_3$. In this reaction, the bicarbonate formed is soluble and is carried away by the underground circulating water. The sites for the dissolution in limestones are the weak zones like joints, fractures and pore spaces. These zones expand with time due to continued solubility, ultimately resulting in the formation of solution cavities (caverns). The acidity of water and the velocity of the circulating waters control the size of these cavities. For the formation of caves of large dimension, the time span involved is of several thousands of years. In the caves, the evidences of solution and the enrichment of carbonate ions in water can be seen (Fig.10.20) from the formation of stalactites (deposition of carbonate in the form of hanging carbonate precipitate) from the leaking waters on the cave roofs and also the growth of these deposits from the leaked water falling on the floor of the cave (stalagmites).

Fig. 10.20 Stalactites and stalagmites in limestone (Courtesy: U.S.Geological Survey Photo : K.Flewelling)

The underground sinkholes in the limestone are often large in dimensions, ranging up to few km in their extent. These sink holes are favourable zones for groundwater storage. The resulting topography in the limestone terrain due to solution process is known as the karst topography. These sink holes are visible on the surface as circular or oval depressions of varying dimensions. In aerial photographs and also on the remotely sensed imageries, the sinkholes show up prominently. Sinkholes in the limestone often coalesce and enlarge in extent.

In limestone terrains, the collapse of the roofs of underground cavities result in the development of sink holes within the limestone of Chattisgarh Basin, as evidenced on the imagery.

Subsidence caused by human activities :

2. ***Underground mining :*** Very common in coalmines, the subsidence of the ground is due to collapse of the roof of the mine. When coal is taken out during mining, enough supports in the mine opening are to be provided to keep the roof in place. However, often due to inadequate supports or excessive mining of coal leaving few pillars for support, results in the roof collapse. This is manifested on the ground surface directly above these mining areas in the form of sinking or subsidence of the ground. In addition, underground fires taking place in some coal fields ultimately result in the ground subsidence. Fires in coal mines once raged continue for several years. Within the Jharia coal fields alone, 70 mine fires encompassing about 17 sq km have been reported to be continuing for considerable period of time resulting in burning of several million tonnes of coking coal so far.

   To locate mine fires, thermal infrared remote sensing is an appropriate and cost-effective method. This tool has been effectively used for the coal mines of Jharia coalfields (Reddy, Dangwal and Bhattacharya, 1998). Aerial predawn Thermal Infra-red (TIR) surveys were carried out at two different periods (March 1989 and April 1995) and on the basis of the data interpretation, the fire areas and direction of fire movement have been monitored (Fig. 10.21).

   Subsidence in the Jharia coalfields in various localities has been common. For example, in Chathali Kuli and Hussainabad areas alone, subsidence has taken place in 92 mines covering an area of about 450 sq km. In 1996, several houses have been reported to be damaged. In these areas, a total of 36 incidences were reported.

3. ***Fluid withdrawal in the subsurface :*** Excessive withdrawal of groundwater from the subsurface results in subsidence. The process involved in known as hydro-compaction, which takes place by dewatering of the pore spaces in the geological material.

   Another process that is associated with dewatering is the oxidation above the water level in marshy or waterlogged areas. Once the groundwater level is lowered, the newly exposed

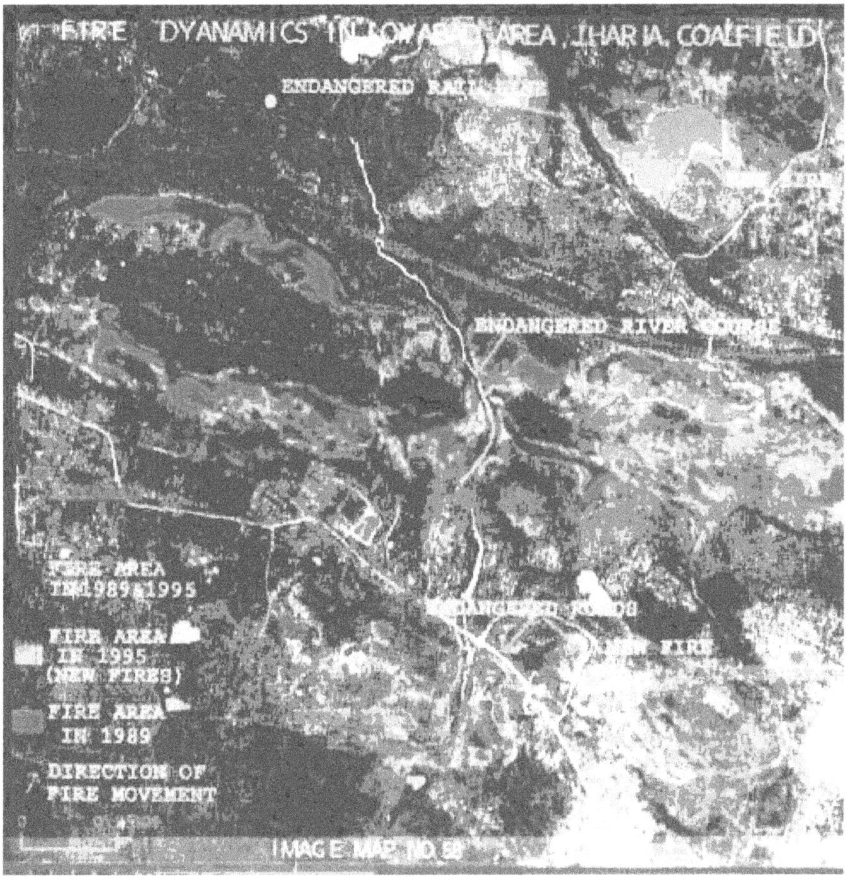

Fig. 10.21 Delineation of underground mine fires in the Loyabad area of Jharia coalfield using thermal infrared remote sensing (Reddy, Dangwal and Bhattacharya, 1998).

material above groundwater table gets oxidized due to the air in the voids. This process induces volume changes in the material, ultimately causing subsidence. In oil and gas fields, withdrawal of the fluids also results in subsidence. For example, several instances of subsidence have been reported from the oil fields in the U.S.A.

11

Environmental Geology

Earth's resources and environment are constantly in interaction with human activities. Since natural resources are finite, a judicious usage is necessary. With increasing demands, environmental degradation is likely. While prevention of pollution is imperative, assessment and remedial measures also are equally important to protect, preserve and optimize the environmental benefits. Environmental geology thus deals with the relation of geology and human activities to environmental exploitation. Natural hazards, although strictly a part of environmental geology, have already been dealt in chapter 10. This chapter deals with environmental degradation as related to human activities.

Groundwater Pollution

Groundwater is the primary source of drinking water and irrigation needs in India, especially on the rural front. Even in towns and cities, it supplements to a large extent the surface water supply for domestic consumption. With the increase in population and consequent demand, there is an excess of withdrawal and that too in an unplanned manner. This situation has resulted in the deterioration of water quality.

Excess of withdrawal affects the groundwater quality in more than one way. Besides the increase in the concentration of the pollutants, the salt water interface gets diffused into the saturated fresh water zone of aquifer resulting in the water turning brackish and sometimes even saline making it unfit for drinking or for agricultural purposes.

Groundwater quality in several canal-command areas has also been reported to have deteriorated. This is typical of waterlogged areas. Waterlogging is common in cases where there is enough infiltration to the groundwater from irrigation canals and the withdrawal is less. Waterlogging is defined by the Central Water Commission as the situation where the water table is less than 1.4 m from the surface. Excessive use of fertilizers and pesticides aggravates the problem, as the groundwater continuously gets richer in these constituents.

Fertilizers and pesticides play an important role in affecting groundwater quality. In India, most common pesticides used are DDT, BHC, endosulfan, aldrin and malathion. High organic content in soil dissolves pesticides which ultimately enter groundwater. The pH and temperature of soil influence leaching of arsenic. Discharge of industrial and domestic effluents, untreated or insufficiently treated, is the principal sources of pollution. About 70% of India's rivers and streams are polluted due to the direct discharge of industrial and domestic effluents. For example, all along the course of Ganga master stream and its tributaries in the Gangetic Plains, the effluents are continuously being discharged from several cities and towns located on its banks. If this water contributes to the groundwater recharge, the groundwater quality is affected.

Pollution from all these sources has resulted in the increase of the levels of flouride, arsenic, nitrates, cyanide, chlorides, chromium, sodium and the total dissolved solids beyond the permissible limits set up by the World Health Organization. This has an impact on the health of living population and often on agriculture.

1. *Arsenic in Groundwater*

This is of enormous concern in India in West Bengal, parts of Bihar. Arsenic pollution in groundwaters of Chandigarh and its vicinity, although not yet in alarming proportions, is a matter of concern. Arsenic pollution in the neighbouring countries – Nepal and Bangla Desh (Fig. 11.1) is quite extensive. All these three countries have the groundwater contribution from the rivers emanating from the Himalayan terrain and the aquifers are interconnected at different depths. In Bangla Desh, three aquifer groups are identified – the upper or composite aquifer closer to the surface, the main aquifer between 6 m and 83 m of depth and the deep aquifer. Arsenic contamination is found to be in the groundwater of the top two aquifers. These are hydraulically interconnected. Spread of pollutants through circulating subsurface water might be due to this reason.

Published reports by research workers in India and Bangla Desh indicate three possible mechanisms for the release of arsenic into the groundwater.

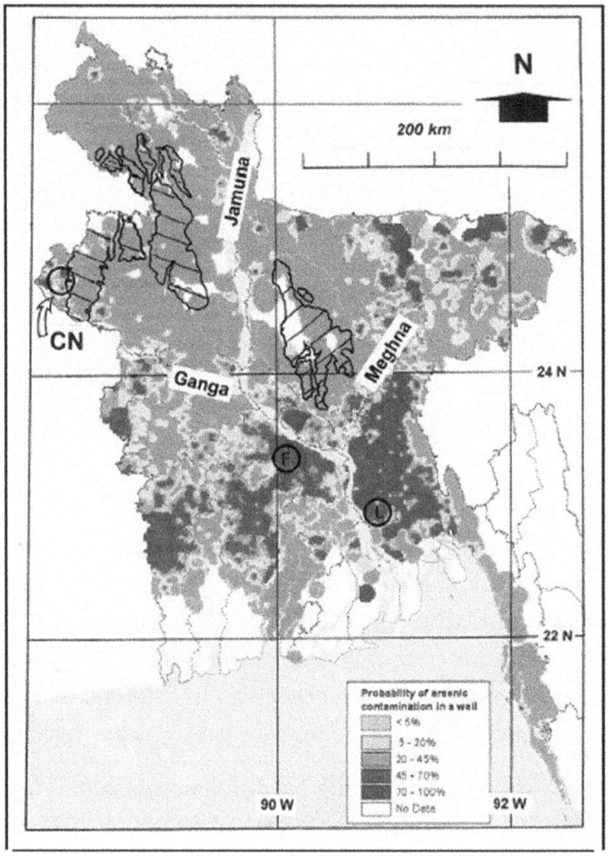

Fig. 11.1 Extent of arsenic pollution in groundwater in Bangla Desh (Mc Arthur, Ravenscroft, Safiullah and Thirlwall, 2001)

1. The pyrite in the alluvial sediments is arsenical in its composition. With the withdrawal of groundwater, consequent invasion of atmospheric oxygen into the pore space of the drained aquifer region causes oxidation of the arsenical pyrite in that zone. The products in the reaction are soluble and released arsenic enriches the groundwater.
2. Dissolution of FeO.OH due to reduction releases arsenic ions into the groundwater. The reduction process has also been thought by some workers to be primarily contributed by the buried peat deposits.
3. Excessive applications of phosphatic fertilizers are thought to be responsible for the release of arsenic by competitive exchange of sorbed arsenic with the phosphate of the fertilizer.

The permissible level of arsenic in drinking water is specified by the World Health Organization as 50 mg / liter. It has been reported that in 42 districts of Bangladesh (covering 492 villages over an area of about 92,100 sq km) and in 9 districts of West Bengal (in 985 villages over an area of about 38,865 sq km), this level of arsenic has exceeded resulting in serious health hazards. The districts affected in West Bengal are Malda, Murshidabad, Nadia, N. - and S. - 24 Paraganas, Burdwan, Howrah and Hooghly.

In Bhojpur District of Bihar, in the Semria Ojha Patti region covered by the villages – Gosaipur, Gopalpur, Ratanpura, Pandeypur and Barisawan, arsenic pollution in groundwater exceeding the permissible level has been reported.

As already indicated, the area around Chandigarh has also been identified to be prone to arsenic pollution.

Arsenic Hazards : Beyond the permissible levels, arsenic in drinking water leads to damage in the liver and nervous system as well as the tumors in the liver. Skin lesions and high prevalence of neuropathy are also common.

2. *Flouride in Groundwater*

Flouride concentration in groundwater is a threat to health in many states in India. As many as 15 out of the 32 states in India are reported to be endemic to flourosis, a disease caused by consuming drinking water containing high levels of fluoride exceeding the permissible maximum safe limit of 1.5 ppm /l set by the World Health Organization. Flouride hazards in humans depend on the level of flouride in the drinking water. While a concentration of about 1.5 to 2.0 ppm /l causes staining of teeth, dental caries and reduction of dental enamel due to osteoflourosis, higher concentrations affect bones. With the increase in concentration, joints in the human body become stiff and brittle. Deformation of bones is a common flouride-induced ailment. The leg bones in particular are affected by the induced osteomalcia, resulting in extreme crippling of the person affected. In India, the flouride content in water in many of the affected regions varies between 2.8 to 12.00 ppm /l.

The flourine is contributed primarily by flouride minerals such as flourite (CaF_2) and flour-apatite ($(Ca_3PO_4)F_3$) which are accessories in pegmatites, granites and gneisses. Phosphatic nodules in the Cretaceous carbonates have also been reported to contain high concentrations of flouride. Thus the flouride hazard is largely confined to the crystalline hard rock terrains and the sediments derived from the same.

Several areas in Andhra Pradesh, Karnataka, Kerala and Tamil Nadu in S.India are reported to be affected by flourosis. The affected areas in A.P. are located in Nalgonda, Ranga Reddy, Medak, Krishna, Anantapur and Karimnagar districts. Maximum concentration has been reported as 28.9 ppm/l in Nalgonda while in all the other sites, the concentration has been cited to be up to 12.0 ppm /l. In Karnataka, the affected areas are situated in Bangalore (Thimmapura village), Kolar, Gulbarga, Raichur, Chitradurga and Tumkur districts. Chengalput and Madurai districts in Tamil Nadu as also Palghat district in Kerala are affected.

Incidence of flourosis and flouride-induced ailments are also reported from the following districts in the respective states as indicated:

Assam : Karbi - Anglong
Jharkhand : Daltonganj
Bihar : Rajauli subdivision
Gujarat : Amreli, Banaskanta, Kutch
Haryana : Hissar, Kaithal, Gurgoan

Orissa	:	Bolangir, Kalahandi
Punjab	:	Amritsar, Bhatinda, Faridkot, Ludhiana, Sangrur
Rajasthan	:	Pali, Sirohi, Nagraur, Bikaner, Ajmer, Jaipur
U.P.	:	Unnao, Mathura, Aligarh, Meerut, Rai Barelli, Sultanpur, Jaunpur, Etawah

Several alternate technologies are available for reducing or eliminating the flouride concentration in drinking water. These include treatment using activated alumina, bone char, mixture of aluminum and lime, ion-exchange resins as well as adopting reverse osmosis. The technique using activated alumina is known as Prasanthi technology (developed at Prasanti Nilayam) while the one involving alumina and lime mixes is called Nalgonda Technology (developed by NEERI).

3. *Pollutants from Industrial Effluents :*

Concentration of various ions in groundwater has been evidenced in different regions in India,mostly thought to be contributed by industrial effluents discharged on to the surface and let into the streams and rivers. Few examples, as reported, are cited below :

In Tiruppur (Tamil Nadu), although several existing textile-processing units have effluent treatment plants on a combined basis, a few other processing units still lack in having efficient treatment plants. The effluents, after treatment, released into the adjacent Noyyal river (a tributary of Cauvery) are reported to be containing total dissolved solids far exceeding the stipulated permissible limit. The dissolved solids appear to be rich in sulphates and chlorides.

The soil and groundwater in the region are rich in these pollutants. This has resulted in the salinity of the groundwater with pH around 8.5. The TDS in groundwater has been reported to be greater than 3000 mg/l.

In Punjab, industrial effluents from the wool industry in and around Ludhiana with high concentrations of cyanide (up to 7.0 mg/l), chromium (around 2.5 mg/l), arsenic (up to 0.04 mg/l) and lead (around 0.05 mg/l) are reported to pollute the aquifers in the region. Groundwater in Ropar district has indicated concentrations of flouride (>1.5 mg/l), high TDS (around 8690 mg/l) and high total hardness (1616 mg/l). This has been localized in two areas and has been attributed to discharge of waste from polluting industries.

In Tamil Nadu, in and around Dindigul, tannery wastes have been reported to be responsible for high concentrations of Na, Ca, Cl, SO_4 and NO_3 in the groundwater.

Tannery sludges used in agriculture have been thought to have caused high concentration of chromium (up to about 6%) in the soils and groundwater of Vellore area in Tamil Nadu. Tannery wastes discharged into the Ganga from the processing units in Kanpur and Unnao districts of U.P. are rich in various metal ions.

4. *Pollution through Fertilizers*

Excessive use of fertilizers and waterlogging conditions contribute to groundwater pollution. For example, groundwater in the N. Krishna delta in A.P. has been found to have high nitrate levels, attributed to excessive use of nitrogenous fertilizers, insecticides and pesticides.

Similarly deterioration of groundwater quality has been related to extensive use of fertilizers in the Sangamner area in Ahmednagar district of Maharashtra. Groundwater in Amritsar City is found to have high TDS, total hardness, nitrate and high pH. Such cases exist in several parts of the country. Groundwater salinity is a common feature in the Bhakra Canal system area where waterlogging coupled with use of fertilizers is dominant.

High concentration of nitrate in drinking water is hazardous for health as it causes methaemoglobinaemia (commonly known as blue baby syndrome) in infants. Human cancer has also been related to nitrate through the formation of extremely powerful carcinogenic agents such as nitrosamides and nitrosamines.

5. *Groundwater Salinity*

Groundwater salinity, besides being due to waterlogging and extensive use of fertilizers, can also be due to excessive and indiscriminate withdrawal of fresh water resulting in the encroachment of salt-water interface. This is common in coastal regions. For example, groundwater salinity in Puri, Cuttack and Balasore districts in Orissa, Haldia and 24 Paraganas in West Bengal , and in Kutch , Banaskantha and Surat in Gujarat has been attributed to this cause.

Groundwater salinity in the Gangetic aquifers in some parts of U.P. has been reported to be due to local enrichment of concretionary nodules (locally called kankar) within some aquifer layers in selected spots such as in Kanpur, Mathura, Raibareilly and Sultanpur.

Environmental Impact of Water Impoundment

Construction of dams with large reservoirs has considerable impact on the environment. Besides the changes in the water flow and groundwater conditions, changes in the sediment transport and the delta formation are also modified. The ecological habitats and human settlements get modified. Further, the impounded water in the reservoir is often a threat to the instability of the region. Details are as follows :

1. ***Reservoir-induced Seismicity :*** Impoundment of water has brought in crustal instability in several cases all over the World. The earliest reported case of reservoir-induced earthquakes was at the Hoover dam in the U.S.A. Some of the major reservoir-induced earthquakes are indicated in Table 11.1.

Table 11.1 Major reservoir-induced earthquakes

Dam	Year	Magnitude	Year of Impoundment
Koyna (India)	1967	6.3	1962
Kariba (Zimbabwe)	1963	6.2	1958
Kremasta (Greece)	1966	6.2	1965
Xinfengjiang (China)	1962	6.1	1959
Hoover (U.S.A.)	1939	5.0	1935

Koyna dam (Fig. 11.2), located in the basaltic terrain is a classic example. It is a concrete gravity dam (85 m high and 800 m long). The reservoir (Shivajisagar lake), located in the watershed region of the Krishna river, has a catchment area of 872 sq. km and water capacity of 2.8 bln cu.m.

Fig. 11.2 Koyna dam (Photo: M.N.Kulkarni).

The reservoir-filling started in 1962 and has reached its maximum level in 1967. Earthquake tremors started in the Koyna region since 1963 and their frequency started increasing over the years. The hypocentres of these shocks, as evidenced from the data, have clustered around the reservoir lake at very shallow depth. In 1967, prior to the major earthquake, five shocks were recorded, of which the one in September was of 5.5 M. The major shock (6.5 m) was in 1967 and the focal depth has been estimated as 8 to 10 km. This shock has been related to the Koyna river valley fault (Donichiwada fault). This earthquake was followed by several aftershocks. The shocks were related to the increase in the reservoir level during the monsoon months. In the Warna reservoir located around 20 km south of Koyna, with impoundment of water beginning in 1986, tremors started appearing from 1993. Between 1993 and 1995 alone, more than 1200 shocks were recorded. The earthquakes in December 1993 and February 1994 were of 5.0 and 5.4 magnitudes respectively. Based on the work done by the National Geophysical Research Institute, two epicentral regions, one passing through Koyna and the other through Warna reservoir, were established. Till date, at least 100 shocks of magnitude 4.0 or more have occurred (with at least 10 of them with 5.0 M). This type of seismic activity that continues for several years after the reservoir-filling is known as protracted seismicity (Talwani, 2000). For monitoring the water levels in the Deccan Trap terrain in the vicinity of the Koyna dam, 21 bore wells, of depth ranging from 90 to 250 m, have been drilled and installed with pressure transducers. In 6 to7 of them, the water level changes related well with co-seismic activity (Chadha and Kuempel, 2003).

Relocation of more than 300 earthquakes that occurred between 1963 and 1995 in the Koyna-Warna region using several related data has been carried out (Talwani, 2003). The spacial patterns were then integrated with all the available geological, geophysical, geomorphic, neotectonic and lake level data including the observations after the Koyna earthquake of 1967. On the basis of all these aspects, the area between the Koyna and

Warna rivers has been divided into several seismogenic crustal blocks, underlain by a fracture zone filled with water. The larger earthquake events conform to this zone. It was also established in this study that two faults occur – the Koyna river fault zone and the Patan fault (Fig. 11.3). In fact, the earthquake of February 1994 in this region is inferred to coincide with the intersection of these two faults.

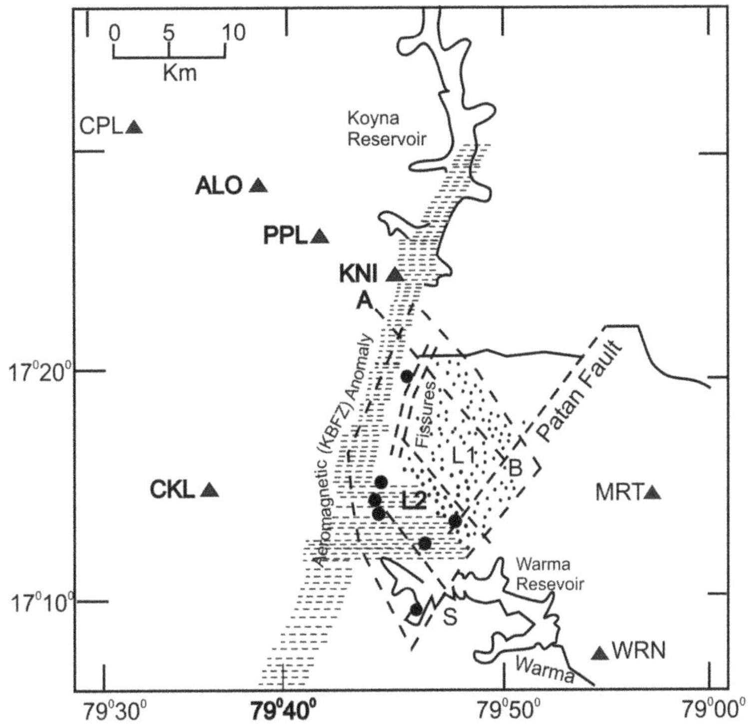

Fig. 11.3 Seismotectonic features in the Koyna Region (Talwani, 2003) (KRFZ : Koyna River Fault Zone, L_1 and L_2 : Lineaments)

Extensive network of G.P.S. (Global Positioning System) stations have also been setup in the Koyna dam region to monitor the deformations of the dam with fluctuations in the water level (Manake and Kulkarni,2002). In all, 34 stations (Fig. 11.4) were located in the vicinity of the dam. Seven of these were on the top of the dam (a centrally- located base station with three stations on either side of it) while three more on the pier and three on the dam base. One station was set up at a place 8 km away N. of the dam (Fig. 11.4). In addition, 8 stations were along the fault line on the downstream side of the dam along with 5 deformation stations and 2 safety stations. On the reservoir, two more stations were located. From the observed data, effect of the plate motion on the dam deflections was eliminated on the basis of data collected from the International GPS station located at Bangalore. Analysis of the reduced data thus conforms to the effect of water table fluctuations due the changing water levels during the post-monsoon and pre-monsoon periods.

Observations over two years have revealed a significant uniform pattern of the deformation as related to the fluctuations in the water level. As the reservoir is full with the highest water level in September, the dam indicated a shift to the downstream direction while it was to the upstream side when the water level was low in month of May (Fig. 11.5). Correlation between the dam structure and fault line has also been reported .

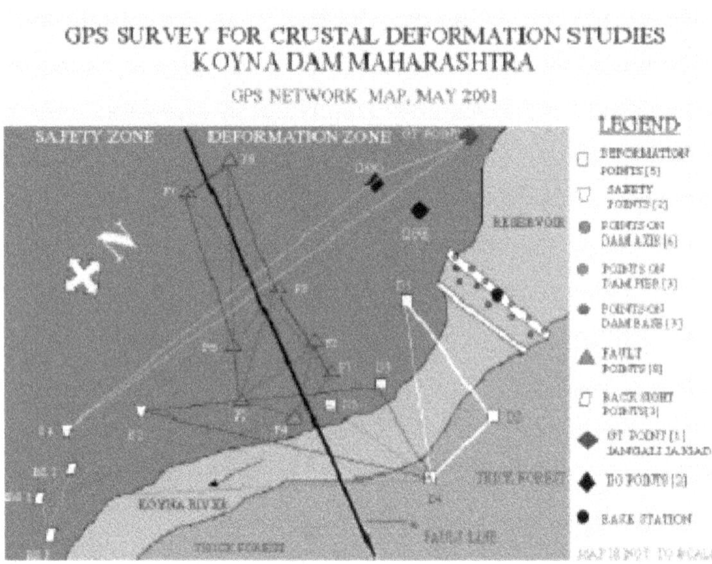

Fig. 11.4 Location of G.P.S. stations in Koyna area (Manake and Kulkarni, 2002).

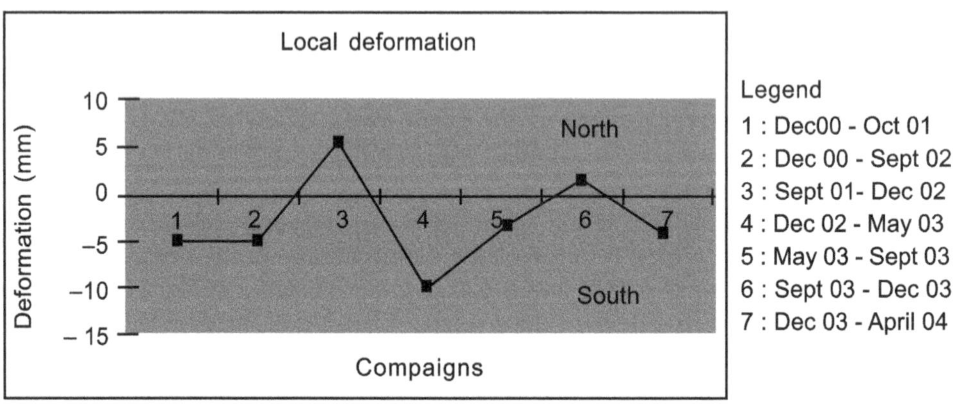

Fig. 11.5 Seasonal deformations of the Koyna dam (Kulkarni et al, 2004).

In Peninsular India, reservoir-induced seismicity has also been reported at six other dam sites namely Idukki (3.5 M), Sharavathy (2.0M), Ukai (3.0 M), Mangalam (3.0 M), Parambikulam (3.0 M) and Kinnerasani (5.3 M).

2. ***Landslides :*** Reservoirs located in geologically unsound river valleys trigger landslides due to water level fluctuations in them. A classic case is that of Vaiont dam located on the Vaiont river, located near Belluno (about 100 km N. of Venice) in N.E Italy. The dam, second highest in the World, is a concrete arch dam of 265 m height located in a synclinal valley. The geological sequence consists of Middle Jurassic limestones at the base of the dam, overlain by Upper Jurassic and Cretaceous limestones interbedded with shales. The formations are jointed and weak. Immediately after the reservoir was filled in 1960, a slide of smaller dimension occurred due to the detachment of a block resulting in the sliding of about 70,000 cu m of material into the reservoir lake. This was attributed to the activation of an ancient slide along the slope of the valley. Saturation of clay material with groundwater was thought to be responsible for the slide As a solution to the problem, drainage and bypass tunnels were constructed to regulate the drainage. Reservoir levels were also contained through several cycles of drainage to control the creep movements. In 1963 (October), a huge mass of rock and earth material (of around 240 M cu m) detached from an area of 2.0 km × 1.5 km slid down into the reservoir lake within around 50 seconds and blocked it completely (Fig. 11.6). Due to this massive filling, a wave of 70 m height generated has flown very high over the dam to the down stream side and the flooding has caused destruction of four villages (Longarone, Villanova, Rivalta and Fae) resulting in loss of around 3000 persons. The slip surface of this slide was within the clay layers and the cause of the slide was attributed to the increase in pore pressures, reduction of effective normal strength with consequent decrease of shear resistance, caused by the increase in the reservoir water level. However, no damage was reported to the dam.

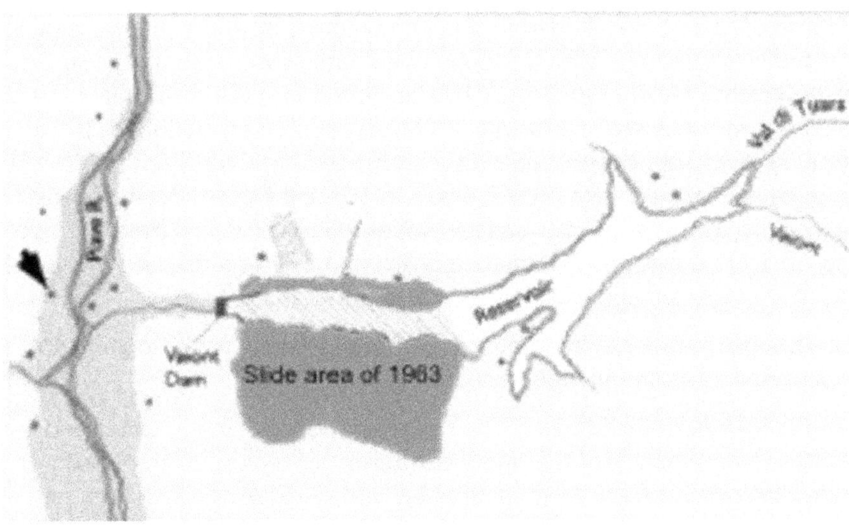

Fig. 11.6 Sketch showing major slide area in Vaiont dam region.

3. ***Changes in Water Quality :*** The sediment load brings in compositional changes in the water. Depending on the depth, the temperature changes and the water is stratified, each layer with a particular composition. In the hard rock areas in S. India, the impounded water in several instances has been found to have flouride content. For example, in the Nagarjunasagar reservoir, increase in the concentration of flouride has been observed since the functioning of the reservoir.

4. ***Downstream Changes :*** Diversion of water causes sometimes drought situation downstream. For example, the Idukki project involved diversion of Upper Periyar water to Muvattupuzha river which has caused drought in summer in the downstream areas besides reducing fresh water availability to the industries located near the mouth of the river. Impoundment of water and diversion have also led to the salt water intrusion where fresh water was available earlier. This is due to the disturbance of the dynamic equilibrium of the fresh water with estuarine tidal cycle.

5. ***Water-logging :*** Common in canal-command areas, water logging and salinity are inter-related. The main causes for the same include higher seepage losses, excessive use of irrigation water, presence of hard saline patches below, poor drainage and porosity of the soil and reduced drafting of the groundwater. The groundwater level in the waterlogged areas is close to the surface (around 2.0 m or less). Due to the capillary rise and evaporation, the salts are brought close to the surface and within the root zone of the vegetation and crops with the result the area is turned into infertile one. It has been reported that in India, about 10 M ha of cultivated land has been affected. In Punjab alone, around 0.1 M ha are categorized as being waterlogged. The other states with considerable waterlogging areas include West Bengal, Haryana and Karnataka.

   In the Nagarjunasagar command area in Andhra Pradesh, the draft from about 6000 wells is reported to have been reduced drastically or stopped with the release of water from the project in 1967. Added to this, the seepage from the unlined canals exists. Most of the canals are unlined. Consequently water table rose closer to the surface (1m to 3m) from the ground level. More than 100 observation wells to monitor the fluctuation of water level in an area of around 1.17 ha have been installed 1975 by the State Groundwater Organization. Test data indicated that between 1975 and 1978, the waterlogging increased from 35% to 68%. In 1978, the groundwater levels were reported to be around 1.4 m from the surface. Investigations by the National Remote Sensing Agency in 1996 have revealed an area of around 1710 ha affected by waterlogging. The extent of salt-affected soils delineated was around 42,800 ha.

   In the Thungabhadra and Sriramsagar command areas, water logging and salinity have been reported since the inception of the projects.

6. ***Loss of Ecological Habitats :*** The flora and fauna on the upstream side are destroyed due to the submergence of the areas in a reservoir region. In the Idukki project, reports indicate that around 6475 ha of evergreen and deciduous forest area was submerged.

7. ***Dislocation of Human Settlements :*** Displaced population from the reservoir area and their resettlement is a common feature. For many of the hydroelectric projects, this operation involves social and legal measures.

Impact of Mining on Environment

Mining operations often cause drastic changes of the landscape. Open-pit mining results in deforestation. There is always a danger of pollution of the surface and groundwater with the mine waste. Iron ore mining in Goa forms a good example. The ore has lot of fine material, which contributes to the high-suspended particulate matter in air. Besides this, the wastewater from the mines and the waste dumps has considerable concentration of the fine red material. The tailings from the beneficiation plants also contribute to pollution. About 35% of the total area (3350 ha) under mining is reported to be covered with mine dumps. Iron mining in Goa has affected the biodiversity. Productivity of agricultural lands has also been impaired.

In the underground mining operations, if the supports are not adequate within the excavated region, sudden collapse of the roof causes considerable damage and loss of life. As explained earlier, subsidence of the surface is quite common in coal mines. Details regarding subsidence are presented in the chapter 10.

Underground flooding occurs in mining operations either due to the breaking of barrier between the working mine and a waterlogged old working or due to sudden tapping of the water under considerable piezometric head. Flooding from surface water is also one of the causes. For example, in the collieries of Jharia and Raniganj coal fields in Jharkhand State and in the Singareni collieries of Andhra Pradesh, underground flooding led to loss of lives. In 1975, in Chasnala colliery (Jharia coal field) disaster, 375 lives were lost. The sudden flooding was attributed to the connecting of the working mine with a waterlogged neighbouring old working. Some of the major cases of flooding in coalmines in India are listed below :

1975	Chasnala colliery	(Jharia C.F.)	375 lives	Connection to old working
1994	New Kenda colliery	(")	55 "	" "
1995	Gaslitand colliery	(")	64 "	Breach of embankment of nearby water course
2001	Bagdihi colliery	(")	31 "	From adjoining colliery
1953	Majri colliery	(Chandrapura C.F.)	11 "	Inundation from Wardha R.
1956	Burra Demo colliery	(Raniganj C.F.)	28 "	Connection to old working
1970	C. Sounda colliery	(S. Karanpura C.F .)	10 "	Flooding from surface
2003	Godavari khani	(Singareni collieries)	17 "	(June)
2003	("	")	15 "	(Oct)

Annexure-I

Geological Time Scale

The concept of geological time has been evolved for the convenience of assigning a specific time period for any group of rocks. Comparision of two groups of rocks in terms of their relative ages also becomes easier once the the time scale is formulated. With this in view the entire period of time (about 4500Ma) since the evolution of Earth has been subdivided into Eras, Periods and epochs. A standard internationally accepted terminology has been established. This is indicated in the Table A-1.

For establishing the geological succession (older to younger formations in a chronological sequence) and to assign particular ages to the same, several approaches of dating are adopted. These can be broadly divided into dating on a relative time basis or absolute time basis.

Relative dating : It enables comparision of rock formations or groups and evaluating their relative position in a geological sequence. For this purpose the following aspects are considered:

1. Fossil succession (both fauna and flora)
2. Cross-cutting (intrusive) relations
3. Inclusions in rocks and minerals
4. Superposition of formations (unconformities)

Different fauna and flora have been prevalent during specific periods of geologic time. Hence a formation containing specific fossil type is considered to belong to that time. Sinilarly, if a rock type exhibits cross-cutting nature with another, it is considered later in age of its formation. The same principle applies to inclusions of rocks or minerals in other formations. Unconformities play an important role in establishing the superposition of formations involved. Unconformity signifies a period of erosion or nondeposition after which the subsequent formations are deposited.Within this intermittent period, if there are structural disturbance leading to tilting (development of dip), folding or faulting of the formations, the earlier deposited formations exhibit these features which will be absent in the later series. All these help in arrangement of the series of formations involved into a relative time sequence.

Absolute dating : This involves estimation of absolute age for any formation. This involves radiometric determination of age in terms of absolute time. For this radioactive isotopes are used. These are unstable and disintegrate into their respective daughter products. The rate of this decay is known as half life. Half-life ($t_{1/2}$) is the time by which half the amount of original parent isotope has disintegrated. Since the geological ages are in the order of millions of years, radioactive isotopes with long half life are to be utilized. The following ate commonly adopted isotopes :

Parent isotope	Half-life (million yrs)	Daughter product
U^{238}	4500	Pb^{206}
U^{235}	713	Pb^{207}
Th^{232}	13900	Pb^{208}
Rb^{87}	50000	Sr^{87}
K^{40}	1500	Ar^{40}

The expression involved for age estimation is

$$D = P(e^{\lambda t} - 1)$$

where D = the amount of daughter product

P = the amount of parent isotope

λ = decay constant

t = time period for decay (age)

The relation between decay constant (λ) and half life ($t_{1/2}$) can be worked out as follows :

If N_0 represents original number of parent isotope and N is the number of atoms after a period of t,

$$N/N_0 = e^{-\lambda t}$$

At $t = t_{1/2}$ (half-life), $N/N_0 = ½$

Hence, $½ = e^{-\lambda t_{1/2}}$

Thus the value of λ can be estimated from the half-life of the isotope used in age estimations. Accurate age determinations can only be possible if the amounts of parent and daughter isotopes are determined accurately. It is based on the assumption that the sample material analyzsed did not contain any lead originally in its composition.

The standard geological time scale along with the faunal and floral distribution and distribution of typical geological groups of formations is presented in Table A-1.

Table A-1

Era	*Period*		*Ma ago*	*Fauna*	*Flora*	*Typical Formations*
Cenozoic	Quaternary	Recent				Recent Alluvium
		Pleistocene				Older Alluvium
			1.6			
	Tertiary	Pliocene				Siwalik formations
		Miocene		Mammals		
		Oligocene				Jaintia & Barail formations (Asam)
		Eocene				
		Paleocene				
			66.4			Deccan Trap basalts
Mesozoic		Cretaceous			First development of flowering plants	Tiruchinapalli / Narmada Valley Cretaceous formations
		Jurassic		Reptiles		Jurassic System (Kutch)
		Triassic				
			245			
Paleozoic		Permian		Amphibians	Plant life (coal-yielding)	Gondwana formations
		Carboniferous				
		Devonian		Fish	Earliest forests	
		Silurian				
		Ordovician		Invertebrates		
		Cambrian				
			570			
Precambrian		Proterozoic		Primitive forms of life		Cuddapah / Vindhyan Groups
			2500			
		Archaen		No life		Dharwar / Aravalli Systems
			4600			

Annexure-II

Internal Structure of Earth

Earth is a differentiated planet. During cooling and consolidating, differentiation according to the density and specific gravity has taken place. This has resulted in heavier material existing at increasing depth. Earthquake records have provided very valuable information regarding the internal constitution of Earth. The P-wave and S-wave velocities indicated a sudden change

The distinct changes in velocities occur at specific depths. These breaks are designated as the discontinuities. The first discontinuity (called the Mohorovicic discontinuity, named after a Yugoslavian seismologist) occurs at about 70km depth from the surface. This discontuity corresponds to the base of Earth's crust. The thickness of the crust varies at different places but its maximum thickness is around 70km on the continents. The ccontinental crust is lighter and its composition is of grantic material. The thickness of the crust under the oceans however is upto around 8km and it is made up of basaltic material. The oceanic bottom crust is younger in age. Even at the present time, at the diverging plate boundaries, the mid-oceanic ridges have the emerging volcanic material forming the new crust.

The second discontinuity is known as the Gutenberg discontinuity (named after Beno Gutenberg, the famous American seismologist). This break has been located at a depth of 2900 km and it corresponds to the bottom of the Mantle (the Mantle-Core boundary). At this boundary, the S-wave propagation abruptly ends, indicating the liquid nature of the succeeding core material . It may be noted that liquids do not take shear. Detailed investigations have established that the composition of the core corresponds to Fe-Ni metals. The mantle is composed of basic material (of Fe-Mg silicates) at the boundary with the crust, which is increasingly basic (with a drastic reduction of the silica component in the composition) with depth. Thus much of the deeper mantle is peridotitic (ultrabasic) in its composition. Subsequent investigations have further revealed that all the core is not of liquid nature and the central part of the core (called the inner core) is solid.

A low velocity layer has been identified in the upper part of the mantle from around 100km to 350km of depth. This zone is designated as the asthenosphere. The strength of the rocks is weak there and the material is believed to be more plastic in nature within that zone.Above the asthenosphere, right up to the surface the zone is called lithosphere.

Below the asthenosphere, the zone up to the core boundary is known as the mesosphere.

The different classifications of the Earth's interior are indicated in the Fig. AII-1.

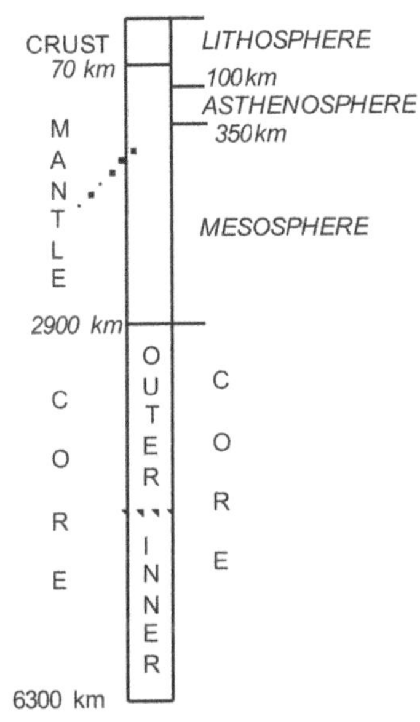

Fig. AII-1 Internal structure of Earth.

Annexure-III

Review Questions

CHAPTER 1

Minerals and Rocks

1. Define “mineral”. How are minerals classified?
2. What can be the maximum number of minerals a rock can be composed of? Why?
3. List the essential rock-forming minerals. From among that list, indicate the minerals common in sedimentary rocks.
4. Why is the Si-O linkage tertrahedral in nature?
5. On what basis are the silicate minerals are classified?. Indicate the classification.
6. What is the difference between “magma” and “lava”? Why are the basic lava flows common in nature compared to silica-rich lava flows?
7. Indicate the sequence of crystallization of minerals (Bowen’s Reaction Series) in igneous rocks. Which mineral is the last one to crystallize in a magma?
8. In what type of igneous rocks can free Al_2O_3 (corrundum) occur in nature? Why?
9. What are the parameters on the basis of which igneous rocks are classified? Indicate the general scheme of classification of these rocks.
10. In which rock type are the columnar joints common? In India, where do these rocks extensively occur? Comment on the process of formation of these joints.

11. Why are volcanic rocks fine-grained? What are vesicles and how do they form? What is amygdaloidal structure in basalts?
12. Differentiate between the following : (Draw neat sketches)
    A. Phaccolith and laccolith B. Dyke and sill C. Lopolith and batholith
13. Indicate the classification of sedimentary rocks.
14. Environments of deposition of sedimentary rocks can be inferred from the nature of the rock types and sedimentary structures present in them. Justify this statement with examples.
15. Sandstones can be classified into three types. What are they? Indicate their basic differences.
16. What do the following rocks signify in terms of their depositional environments:
    A. Boulder bed
    B. Conglomerate
    C. Graywacke type of sandstone
    D. Sandstone-shale sequence without limestone (as in the coal basins)
17. What is cross-bedding and how does it differ from graded bedding? Illustrate your answer with neat sketches.
18. What are ripple marks? Indicate their subdivision into different types. Comment on their significance in terms of nature of their formation.
19. Outline briefly the different sedimentary structures and indicate their importance in the study of sedimentary rocks.
20. What is diagenesis? What are the evidences for the same in sedimentary formations?
21. Pressure, temperature and composition of the pre-existing rocks control the formation of specific metamorphic rocks. Justify this statement with suitable examples.
22. What are the different types of metamorphism? Mention the conditions prevalent in each of them.
23. How are metamorphic rocks classified? Name six metamorphic rocks and indicate the original rock (prior to metamorphism) in each case.
24. Define (a) metamorphic grade and (b) metamorphic facies. Draw the pressure-temperature diagram and locate the regions of different metamorphic facies.
25. Write a short note on suitability of rocks as construction materials
26. Comment on the properties of granite and marble as building and decorative stones. Cite important occurrences.
27. What are alkali-aggregate reactions? Outline, in brief, the characteristics of different types of these reactions.
28. Chlorides and sulphates are considered undesirable in concrete. Justify with reasons
29. Classify the following rock types (in terms of igneous/sedimentary/metamorphic) and indicate important minerals in each of them :
    (a) Granite (b) Limestone
    (c) Khondalite (d) Charnockite
    (e) Gabbro
30. Which is the oldest group of rocks in S. India?. Where do these rocks occur? Comment on the nature of these rock types.

CHAPTER 2

Engineering Properties and Classification of Rock Masses

1. What is a rock mass? What are the advantages of engineering classification of rock masses compared to the conventional classification of rocks?
2. In addition to the engineering classification, conventional geological classification is still beneficial in civil engineering practice. Justify.
3. What are the different engineering properties of importance in site evaluation? Categorize them in terms of field tests and tests in the laboratory on specimens.
4. Draw typical stress-strain curves for granite and clay. What is the essential difference in their deformation behaviour ?.
5. Indicate typical ranges of unconfined compressive strength values of igneous, sedimentary and metamorphic rocks (for four rock types in each of them).
6. What is tangent modulus? How is it estimated from the stress-strain curve?
7. How is uniaxial tensile strength of a rock specimen obtained? Why is it not a direct test as in the case of metallic and ceramic materials?
8. Indicate, on a neat sketch, the theoretical stress distribution pattern in a rock specimen subjected to indirect tensile strength test (Brazilian test).
9. What is the significance of shear strength values in field practice? Using Mohr circle method, how is shear strength estimated? Draw the Mohr diagram to illustrate your explanation.
10. What are the advantages of estimation of dynamic elastic parameters compared to obtaining static values? Give the field procedure and the expressions relating dynamic elastic parameters to the seismic wave velocities (of P- and S- waves).
11. How are rock masses classified on the basis of unconfined compressive strength and tangent modulus (Deer and Miller classification)? Indicate the schemes.
12. What is Rock Quality Designation (RQD)? How is RQD used for the rock mass classification?
13. What is Rock Structure Rating (RSR)? What are the parameters involved in estimation of RSR values of rock masses?
14. List all the six parameters involved in the Rock Mass Rating System (proposed by Bieniawski). How are they used in rock mass classification based on this system?
15. What are the parameters used for the estimation of Rock Quality Index (RQI in Barton's method)? Indicate the equation used and explain the significance of each of the ratios involved.
16. List all the essential rock mass characteristics needed (from both field and laboratory testing) to classify any rock mass on the basis of RSR, RMR and RQI systems of classification.

CHAPTER 3

Weathering and Soil Formation

1. What is mechanical weathering? How is it different from chemical weathering? List the different processes operative in chemical weathering.
2. Arrange rock-forming minerals in the increasing sequence of their susceptibility to weathering. What are the reasons for variation in their durability to weathering?
3. Draw a neat sketch to illustrate weathering as a function of temperature and rainfall.

4. Draw a typical profile of soils in the tropical climate. How does it differ from a profile common in desert climates?
5. What is slake-durability index? Indicate the steps involved in the same.
6. How is a rock mass graded qualitatively in terms of the degree of weathering? List the different grades in the classification.
7. How are ultrasonic velocities used in classifying a weathered rock mass? Indicate the different grades.
8. How is the weathering index computed in terms of quartz to feldspar ratio in a weathered rock mass? Indicate the weathering grade classification based on the same.
9. Draw an idealized soil profile with all the possible soil horizons. Indicate the characteristics of each of the horizons.
10. What are pedocal and pedalfer? Under what conditions each one of these soils are stable?
11. What are differences between saline and alkaline soils? Give Indian examples
12. List the differences between eolian and marine soils in terms of their mode of formation, grain size and sorting.
13. Draw a neat sketch of the plasticity chart used by geotechnical engineers for classifying the soils. Label all the areas on the chart in terms of different soil types.
14. Outline in detail the characteristics of lateritic soils. Draw a typical soil profile for the same. How are these soils different from black-cotton soils? Provide Indian examples for both these soils.
15. What are engineering problems encountered in construction in a black-cotton soil? What is the cause for the problem behaviour in these soils?
16. What are the main parameters on the basis of which soils are classified by the soil scientists (F.A.O. classification)? How is this approach different compared to the classification used by geologists or by the geotechnical engineers?
17. Indicate the chemical reactions involved in hydrolysis. How is hydrolysis different from hydration in weathering? .
18. Draw a neat diagram indicating the variation in weathering from arctic region to equatorial region.
19. List the different types of soils in India. Comment on their distribution.
20. What are changes and involved reactions during the weathering of granite? What is the final product?

CHAPTER 4

Geomorphic Processes and Landforms

1. How do ephemeral streams differ from perennial streams? Provide two Indian examples for each one of these types.
2. Draw a neat diagram and explain the role of different parameters that control the flow in a stream. How is laminar flow different from a turbulent flow?
3. Define the terms: bifurcation ratio, drainage density and length ratio.
4. Mention various drainage types and discuss their important characteristics.

5. On a satellite imagery of a region, dendritic drainage in a region suddenly changes to a simple drainage pattern oriented in a particular direction. What information does this scenario present in terms of the formation characteristics?
6. Comment on the nature of load carried by a stream. What are the parameters controlling the deposition of different types of sediment load in a drainage channel?
7. Draw the Hjulstrom curve and explain the role of threshold conditions for erosion, transportation and deposition of sediment load.
8. Explain with neat sketches the differences between a meandering and a braided channel. What is an entrenched meander?
9. Discuss briefly the types of landforms and drainage characteristics during youthful and maturity stages of a drainage system. Cite Indian examples.
10. Define the terms: peneplain , unpaired terraces and levees. Draw neat sketches.
11. What is an oxbow lake? Draw neat diagrams to explain its development.
12. How are braided bar deposits formed? How do they differ from point bar deposits?
13. What are the different subdivisions in a deltaic region? Draw a neat diagram of a typical deltaic region to indicate the same. To what type does the Ganga-Brahmaputra delta conform?
14. How is the 'celerity' of a sea wave different from its 'wave length'? Under what conditions does the particle geometry vary in a wave?
15. What is wave refraction? Explain its role in the process of headland erosion and formation of bay beach.
16. Discuss the seasonal variations in the beach profiles in a coastal region. Draw sketches.
17. What are the different types of coral reefs? Under what conditions do they form?
18. What is an estuary? How is a positive estuary different from a negative one?
19. What are the different shore protection structures adopted to control coastal erosion? Draw diagrams to indicate the details.
20. Explain the terms: offshore breakwater, sea wall and groin.
21. Discuss the characteristics of glacial erosion and valley development. What recharacteristic features are evidenced in glacial deposits? What is the significant difference between fluvial load and a glacial load?
22. In India, the Gondwana and Siwalik sequences of formations are considered to be of glacial origin. List the supporting evidences for the same.
23. List various types of sand dunes. Outline the differences between them.
24. What is loess? How do its properties pose engineering problems during construction in a loess region.
25. Indicate the different desert landforms and discuss their characteristic features.
26. Chilka lake in Orissa and a lake (Pangong Tso) in the high altitude (around 4000m) in the Leh region in extreme north of our country are both saline in their water quality. Indicate the causes for the salinity in both these cases . Do you visualize the reasons to be different in both cases ?

CHAPTER 5
Crustal Deformation and Geological Structures

1. Under what temperature, pressure and strain rate conditions do the rocks in the crust behave as brittle materials? How do they differ for material to exhibit ductile behaviour?
2. Draw typical stress-strain diagrams for brittle and ductile rocks.
3. Comment on the brittle-ductile transitions within Earth's crust and mantle regions. Draw a sketch to support your answer.
4. Define strike and dip. Can apparent dip be higher in amount compared to true dip? Why?
5. Explain how the dips of formations are estimated from a geological map. List the steps involved.
6. What is an anticline? How do you differentiate a symmetrical anticline from an asymmetrical anticline?
7. Classify folds into different types on the basis of the axial plane orientation.
8. What is a monocline? How is an isoclinal fold different from a monocline?
9. What are plunging folds? Draw neat diagrams of a plunging anticline and a plunging syncline.
10. Define: (a) normal fault, (b) reverse fault and (c) thrust fault. Draw sketches of the same.
11. How are strike-slip faults differentiated from the dip-slip faults? .What types of faults develop at the tectonic plate boundaries in the mid-oceanic regions?
12. Type of a fault developed in nature is a function of the three principal stress directions (major, intermediate and minor stress orientations). Justify this statement with sketches.
13. List ten criteria for the recognition of faults in the field.
14. On the top of Mount Everest, older rocks occur. How do you account for the same? What important geological structures are responsible for this situation?
15. What are joints and how do they differ from faults?
16. Draw a diagram of an anticline and indicate tension, strike and oblique joints as related to the maximum principal stress.
17. What is an unconformity? What does it signify? What is the most common rock type that is present at an unconformity?
18. Define: (a) angular unconformity, (b) disconformity and (c) nonconformity. Draw neat sketches to support your answer. Label the sketches completely.
19. What are the tectonic changes that occur at the divergent plate boundaries? How are these different from those in the convergent plate boundary regions? Illustrate your answer with neat sketches.
20. Mountain-building is related closely to crustal deformation. Justify this statement using Himalayas as an example.
21. Development of Himalayas is through a process different from that for the development of mid-oceanic ridges. Give the details for both these processes.
22. How are geological structures in rocks important in civil engineering practice?

CHAPTER 6

Site Investigations and Ground Improvement

1. How is a topographic map (prepared by the Survey of India) for an area selected? What does the number, present on the right-hand top corner, of the map signify?
2. An engineering geological map is different from a regular geological map. Justify.
3. What are the elements considered in aerial photo interpretation? Outline in detail.
4. What are advantages of remote sensing data obtained through satellites in preparing the geological map of a region? Give examples in support of your answer.
5. Briefly review the nature of data used in mapping, obtained from various IRS satellites. How do they compare with data obtained from other International satellites?
6. GIS is useful in preparing maps such as hydrogeological, engineering geological and landslide hazard zonation maps. Indicate the basic steps involved in the same.
7. What are the essential differences between P-waves and S- waves? How are these waves useful in providing significant information on internal constitution of the Earth?
8. Indicate the procedure for the estimation of dynamic elastic parameters of various formations to depth using seismic refraction technique.
9. What are travel-time curves? How are dipping formations and presence of a fault confirmed on travel-time curves?
10. Outline the procedure or the estimation of thickness of formations using seismic refraction technique.
11. How is Wenner configuration of electrodes different from the Schlumberger configuration in electrical resistivity method? Indicate the formula used for estimation of resistivity using Wenner configuration.
12. What are Type Curves and Auxiliary Curves? How are they useful in interpretation of resistivity field data?
13. What is resistivity logging? Indicate the advantage of a resistivty log compared to a driller's log.
14. What are the various types of drilling methods used in subsurface exploration? Indicate their suitability for different geological formations.
15. How is reverse rotary drilling different from direct rotary drilling? Mention the specific advantages of the reverse rotary method.
16. What type of a drill bit is used in rotary core drilling? Indicate the different sizes of the cores normally obtained through core drilling of geological formations in a project site.
17. What is grouting? List the different purposes for which it is resorted.
18. Discuss in brief the essential features of different grouting methods undertaken for ground reinforcement in Civil Engineering projects. Illustrate your answer with neat sketches.
19. For back-filling of excavated weak zones, how is the depth of grouting decided? Give examples.
20. What is shotcrete? Indicate the basic differences in the mode of application of dry and wet shotcrete.
21. What are the different types of rock bolts used for ground reinforcement? Indicate their specific applicability.

22. What is preloading (pre-compression)? How is it different from dynamic consolidation? Discuss in brief the salient features of both these methods.
23. What is soil stabilization? Indicate the additives commonly used in the soil. How does the strength development take place during chemical stabilization of soils?
24. What is the main role of lime in soil stabilization? Comment on the nature of pozollanic products formed with lime in the soils during stabilization.
25. What are geomembranes? How do they help in ground improvement? Illustrate your answer with suitable sketches.

CHAPTER 7

Hydrogeology

1. Draw a neat sketch illustrating the various components in hydrologic cycle. How is hydrologic cycle useful in Water Resource Engineering?
2. Indicate the water budget equation. Discuss its use in groundwater assessment.
3. What are the various forms of subsurface water? Draw a neat sketch to illustrate their occurrence at different levels.
4. What are water table contour maps? Discuss briefly their utility in groundwater exploration.
5. Define porosity. What is the difference between primary and secondary porosities?. How is permeability different from porosity?
6. Define the term aquifer. How is aquifer different from aquitard , aquifuge and aquiclude? Give one example for each of them.
7. Draw neat sketches to illustrate unconfined aquifer and confined aquifer. What are basic differences in their operational performances for groundwater yield?
8. Is it true that the piezometric surface in the case of a confined aquifer is always below the ground surface like a water table for an unconfined aquifer? Justify your answer with a supporting sketch.
9. What is a leaky aquifer? Give specific reasons for the occurrence of leaky aquifers in nature. Where are these aquifers common in India?
10. Discuss briefly the performance characteristics of different formations as aquifers.
11. What are fence diagrams? How are they prepared? Discuss their importance in groundwater exploration.
12. How is remote sensing useful for groundwater exploration? Indicate briefly the approaches adopted for locating zones with salinity.
13. How is groundwater quality estimated from resistivity logs obtained in bore holes? Indicate the expression involved connecting water quality with the resistivity parameters.
14. How is fresh water – salt water interface established in coastal areas? Indicate the details of the procedure adopted for the same.
15. In a coastal area, 40 units of fresh water accumulation supresses the salt water level by 1 unit. Formulate the equation for this and indicate various parameters involved on a neat sketch.
16. Name three tracers used in groundwater exploration. Discuss the utility of tracers in groundwater exploration.

17. What is storativity of an aquifer? How is it different from permeability? How is storativity estimated in the field?
18. Define the terms: 'radius of influence' and 'drawdown'. How are time-drawdown curves useful in aquifer evaluation?
19. What are type curves? How is a leaky aquifer identified in curve matching of its time-drawdown curve with the type curve?
20. If the diameter of a tube-well is doubled, will the yield from the well be double of its earlier yield? Justify your answer with relevant equations. (Assume that the other parameters in the equation except the well diameter are kept constant).
21. How are springs classified? Draw neat sketches to indicate different types of springs.
22. Under what conditions thermal springs occur? Discuss briefly the distribution of geothermal provinces in India.
23. What is a groundwater province? List the different groundwater provinces of India. Indicate in brief the important features in each of them.

CHAPTER 8

Engineering Geology in Dam Construction

1. What are the different purposes of dam construction? What is the single main purpose served by 90% of the dams in India ?
2. How is an earth dam different from a concrete dam? Draw a sketch to indicate the salient features of an earth dam.
3. What is an arch dam? Draw sketches to illustrate its main features. Cite an Indian example.
4. A buttress dam is preferred even in weak foundation sites. Justify this statement with supporting sketches.
5. What are all the requirements for the selection of a dam site? Discuss briefly the geological considerations for choosing a dam site .Under what considerations an arch dam is preferred?. How are these different in the case of an earth dam?
6. List all the types of exploratory investigations undertaken at different stages during site location for a dam and its reservoir.
7. How are geophysical investigations beneficial in site exploration for a dam? Indicate as to how all he relevant rock mass characteristics are obtained.
8. Grouting plays an important role in the foundation treatments at dam sites. Outline in detail the grouting methods and specific purposes in dam construction.
9. Discuss the seepage problems in different types of dams. How are they countered? Cite specific Indian cases for the same.
10. What are the causes of silting in reservoirs? Indicate the impact of silting on the functioning of a reservoir. How is silting in reservoirs minimized or prevented?
11. Draw neat sketch to indicate the foundation geology of Bhakra dam site. List all the strengthening measures adopted for improving the performance of the formations.
12. Discuss the geology of the Nagarjunasagar dam site in Andhra Pradesh. What are the foundation treatments adopted and why?

13. What type is the Srisailam dam in Andhra Pradesh? What was the need for the preference of this type? Discuss the salient geological features at the dam site. What were the measures adopted to confine soft seams on either side of the dam and also to prevent sliding within the rock formations on the downstream side?
14. Briefly outline the important features of Idukki dam. How is the dam anchored to the embankments?
15. Where is the Hirakud dam located? What are its important features? What was the need for the selection of the specific type for this dam?
16. Where is Tehri dam located? Outline briefly the geology of dam and reservoir foundations.
17. What are the special characteristics of Himalayan rivers that pose challenges to the dams and reservoirs in their catchment areas? Elaborate your answer with Tehri dam as a case study.
18. Indicate the location , type and geological formations at the site of the following dams :

    (a) Hirakud dam (b) Rihand dam (c) Idukki dam (d) Tehri dam (e) Koyna dam
19. What are the possible disadvantages of constructing large dams in the Himalayan region?. Compare the geological, geomorphological and tectonic situation of the Himalayan terrain with those in the Peninsular part of India in regard to dam construction.

CHAPTER 9

Engineering Geology in Tunneling Practice

1. Tunnels serve the needs for various operations. List the same.
2. How are tunnels classified in regard to the competency of the ground and tunnel geometry?
3. What are the field and laboratory investigations undertaken during the planning stage of tunnel location and construction?
4. Review the performance of different geological formations in tunneling operations.
5. Draw suitable sketches and explain the considerations for locating tunnels in folded and faulted formations.
6. What are the different blasting methods adopted in tunneling? What is sequencing in blasting and how does it help in achieving the tunnel geometry?
7. Outline in detail different modes of ground failures encountered in tunnels.
8. What is overbreak in tunneling? How is it different from a pay line?
9. Indicate the mode of stress distribution in tunnels. Draw a suitable sketch.
10. How are tunnel supports chosen on the basis of RSR classification of rock masses?
11. What is equivalent dimension? How is it different from Excavation Support Ratio (ESR)?
12. Indicate the ESR ratios used for various tunnel support categories.
13. How is the length of a rock bolt decided? Indicate the involved expression.
14. What are the salient features of the New Austrian Method of Tunnelling? How are rock masses classified in this method for different tunnel support systems?
15. Outline briefly the blasting sequence and method of excavation adopted for the underground power house in the Sardar Sarovar Project (Gujarat).

16. Briefly review the tunnel types, seepage problems and ground supports as related to geology in the tunnels on the Jammu to Srinagar national highway in J&K state.
17. What are the geological formations encountered in the underground powerhouse cavity at Chibbro in the Yamuna Valley Hydel Project? Indicate the details regarding the blasting sequence and special tunnel supporting system (perfo-bolts and cable anchoring system) adopted.
18. What are the problems encountered and measures adopted during tunneling in soft formations as well as hard rocks?

CHAPTER 10

Geological Hazards

1. How are earthquakes caused? Explain with the help of a neat diagram.
2. What is plate tectonics? How does it help in understanding distribution of earthquakes and volcanoes?
3. What are the crustal changes that take place at the divergent plate margins?
4. Where do you expect the most recent (youngest) crust on the Earth and why?
5. Draw a neat sketch of a typical seduction zone and explain the different features associated with a subduction zone. Cite one example of subduction zone.
6. What are the different types of plate boundaries? Indicate the same with sketches.
7. How is a seismogram different from a seismograph? What are strong-motion recorders? Indicate their use in Civil Engineering.
8. What is the difference between 'focus 'and 'epicenter' of an earthquake ? How is epicenter of an earthquake located? Explain with the help of a sketch.
9. Indicate the classification of earthquakes on the basis of depth of focus.
10. Explain the terms 'intensity' and 'magnitude' of earthquakes. In which region in S.E Asia did the strongest earthquake occur in December 2004 and what was its magnitude?
11. What are tsunamis? How are they different from seiches? Cite an example of a recent tsunami. Explain its origin, mode of propagation and effects.
12. Indicate, in brief, the different levels of intensity in the Modified Mercalli Scale of earthquake intensities.
13. How is magnitude of an earthquake related to the energy involved?
14. What is seismic zoning map? How is it useful in Civil Engineering practice? On a seismic zoning map of India, which is the severe zone (highest zone) and comment on the regions covered by the same.
15. What were the considerations that led to periodic revision of the seismic zoning map of India?
16. List the possible hazards due to the occurrence of earthquakes. Cite examples.
17. What is liquefaction? How does it take place? Indicate the possible types of structural damage due to liquefaction.
18. How does the type of construction influence its damage during an earthquake? Discuss briefly the stability aspects of different structures during an earthquake.
19. List the different types of volcanic hazards. Indicate in brief their essential characteristics.

20. Where is the only active volcanic spot in India located? Comment on the phases of its activity in the past.
21. What is a lahar? Indicate its characteristics.
22. How is a volcanic hazard map prepared?
23. Indicate the classification of earth movements.
24. What are the causes of landslides? How can landslides be prevented?
25. What are the parameters considered in the analysis of a slope failure? Draw a neat diagram to indicate the same.
26. Slopes are considered open systems with periodic inputs and outputs. List these parameters.
27. Where are the landslides most common in the Himalayan terrain? Explain the tectonic reasons for their occurrence. In Uttaranchal, indicate the geology of landslide-prone areas.
28. Discuss briefly the important features of the landslide zones in the Nilgiri region of Tamilnadu.
29. What are the causes of subsidence?
30. What is a landslide hazard zonation map? How is it prepared? Indicate the role of GIS in the same.
31. How is remote sensing useful in the mapping of the areas of underground mine fires in a coal field? Cite an example and provide details.
32. What are stalactites and stalagmites? In what regions are they expected?

CHAPTER 11

Environmental Geology

1. What are the possible sources of groundwater pollution? Cite examples for the same.
2. Arsenic pollution in groundwater is increasing at alarming rates in Bangladesh and in India. Review this situation giving all the relevant details.
3. Outline in detail the three possible mechanisms of arsenic release into groundwater.
4. What is the permissible level of arsenic in groundwater for drinking purposes? Indicate the areas in India reported to have higher level of arsenic than this limit in the groundwater. Briefly indicate the possible sources for arsenic pollution.
5. What are the possible sources of fluoride in groundwater? Discuss briefly the likely hazards of fluoride to human health.
6. Name two technologies available for eliminating flouride from drinking water. Provide their important features.
7. Pollution of groundwater with industrial effluents is causing concern in India. Cite few examples and discuss the same with details.
8. What are the effects of water-logging in canal-command areas? Cite suitable Indian examples.
9. What are the causes for groundwater salinity? Cite Indian examples of tracts with brackish groundwater.
10. What is reservoir-induced seismicity? How can it be confirmed for any reservoir of a dam?
11. Review briefly the effects of seasonal fluctuations in the Koyna reservoir on the crustal stability.

12. Discuss the geology of Koyna and Warna reservoir areas with special reference to seismogenic crustal blocks and geological structures (lineaments and faults)
13. Outline the environmental impacts of impounding water in a reservoir.
14. What are the environmental effects of mining? Cite examples.
15. Sudden flooding in underground mines is often experienced leading to disaster. Cite possible reasons for the same, giving suitable examples from Indian coalfields.
16. Landslides are triggered in reservoir regions of geologically unsound valley slopes. Justify in detail this statement with suitable examples.

Annexures I & II

1. What is a geological time scale? Indicate its purpose.
2. What are the aspects on the basis of which relative dating is carried out? Explain in detail.
3. What is radiometric dating? List the isotopes used along with their half-life values.
4. Explain the principle involved in radiometric age determination. Indicate the steps involved in the estimation of gological age.
5. Give the standard geological time scale.
6. What are the main features of Gondwana formations? Indicate their position on the standard geological time scale. What is their economic importance?
7. Cite an example of an Indian dam site situated on Siwalik formations. To which geological period do the Siwalik formations belong?
8. Towards the end of Cretaceous period, reptiles suddenly disappeared and mammals started appearing. What were the reasons for the same?
9. What are Mohorovicic and Gutenberg discontinuities? Why are they called discontinuities? Indicate the respective depths of their occurrence within the Earth's interior.
10. Draw a diagram to indicate the lithosphere, asthenosphere and mesosphere.
11. The core of the Earth is drastically different from the crust and mantle. Explain.
12. How is the presence of liquid core established? Give details.
13. What type of rocks do you expect in mantle region and why?
14. Can any earthquake be triggered with its focus deep in the mantle region? Justify your answer.
15. Draw a neat sketch to indicate the internal structure of Earth. Label the sketch completely giving the respective depths and nature of composition of the material at different depths.

References

Ahmad, S.M. (1984) : Hydrogeological investigations for augmenting water supply in drought-affected areas of Banda District, U.P., India – Report of CGWB, India.

Badgley, P.C. (1965). Structural and Tectonic Principles. Harper & Row, New York.

Bajpai, V.N. (1989) : Surface and subsurface evidences of neotectonics and aquifer disposition in Gangetic Alluvial Terrain of Kanpur-Unnao region, Uttar Pradesh,India- PHOTONIRVACHAK , Jour. of Ind. Soc. Remote Sensing ,Vol.17, No. 2, 47-53.

Bajpai, V.N. and Gokhale,K.V.G.K. (1986) : Hydrogeomorphic classification of the Marginal Gangetic Alluvial plain in Uttar Pradesh, India using satellite imageries- Jour. Geol. Soc. Of India, Vol.28, July 1986, 9-20.

Barton, N., Lien,R.and Lunde,J (1974) : Engineering classification of rock masses for the design of tunnel support, Rock Mechanics, Vol.6, No. 4, 189-236.

Barton, N., Loset, F., Lien, R. and Lunde, J. (1980) : Application of Q-System in design decisions concerning dimensions and appropriate support for underground installations, Intl Conf. on Subsurface Space, Stockholm, Subsurface Space, Vol.2, 553-561.

Berg, G. V. (1964): The Skopje Yugoslavia Earthquake, American Iron & Steel Inst., New York.

Bhandari, R.K. (1987): Slope stability in fragile Himalayas and strategy for development, Indian Geotechnical Journal, I-81 (special lecture).

Bhandari, R.K. (1999): Natural disaster in Asia and IDNDR stimulus, Proc. XI Asian Regional Conf. on Soil Mechanics and Geotechnical Engg., Seoul.

Bhattacharya, P.K. and Patra, H.P. (1968) : Direct Current Geoelectric Sounding- Principles and Interpretations. Elsevier, Amsterdam, 135pp.

Bieniawski, Z.T. (1976): Rock mass classification in Rock Engineering, Proc. Symp. on Rock Exploration, Cape Town, 1, 97-106.

Bieniawski, Z.Y. (1979) : Tunnel design by rock mass classification, U.S. Army Engineer Waterways Exptl. Stn. Tech Rep. GL-79-19, 71pp.

Bieniawski, Z.T. (1989): Engineering Rock Mass Classification. Wiley, N.Y.

Billings, M.P. (1968): Structural Geology. Prentice Hall of India Pvt., New Delhi.

Bulletin 20, Global Volcanic Network, Museum of Natural History, MRC129, Smithsonian Institution, Washington DC.

Cambefort,H. (1977): The principles and applications of grouting, Quart. Jour. Engg. Geol., Vol. 10, 1977, 57-95.

Chakraborti, D. et al (2003): Arsenic groundwater contamination in Middle Ganga Plains, Bihar, India – A future danger, Environmental Health Perspectives , 111(9).

Chalapathi Rao, R.V. (1975): Nagarjunasagar Project, Andhra Pradesh, Misc. Pub, No.29 (Engg Geology case histories), Pt. I, 62-74.

Chalapathi Rao, R.V. (1972): Geology and foundation treatment of Nagarjunasagar dam, Andhra Pradesh, Proc. Symp. On Geological and Engineering Problems of River Valley Projects (Auden Commemorative Vol.), Indian Soc. of Engg. Geol , 19-32.

Chalapathi Rao, R.V. and Narasimham, C.V.L. (1975) : Srisailam Hydroelectric Project, Andhra Pradesh, Misc Pub. No 29, (Engg. Geol. Case histories), G.S.I 164-178.

Chandrasekharam,D.(2000): Geothermal resources of India . Country update, Proc. World Thermal Energy Congress, Japan, 134-146.

Chandrasekharam,D. , Ayaz Alam, M. and Minissale , A. (2003): Geothermal resource potential of Himachal Pradesh, India, Intl. Geothermal Conference, Reykjavik, Sec.24, 15-20.

Chopra, B.R. (1977): Landslides and other mass movements along roads in Sikkim and West Bengal, Bulletin IAEG, 16, 162-186.

Choubey, V.D. (1998): Landslide hazard assessment and management in Himalayas, Proc. Intl Conf. on Hydropower Development in Himalayas, Shimla, 220-239.

Choubey, V.D. (1989): Stability assessment and remedial measures of landslides in Himalayas, N.E.India, 28th Intl. Geol. Cong, Washington D.C., 288.

Choudhury, U.K. et al (2000): Groundwater arsenic contamination in Bangla Desh and West Bengal, India, Environmental Health Perspectives, Vol. 108, No.5 .

Datta, D, V, and Kaul, M.K. (1976): Arsenic content of drinking water in villages in N. India : A concept in Arsenicosis, Jour Assoc. Physicians, India, 24, 59-04.

Datta, N. P. et al (1972): Detection of a fault by the seismic refraction method in Kopili Hydroelectric Project, India, Geoexploration ,10, 101-108.

Davis, G.H. (1984): Structural Geology of Rocks and Regions. John Wiley & Sons. New York.

Dearman, W.B. (1974): Weathering classification in the characterization of rock for engineering purposes in British practice, Bull. Intl. Assoc. Engg. Geology, No.9, 33-42.

Deere, D.U.(1968): Geological considerations *In* Rock Mechanics in Engineering Practice,. K.G.Stagg and O.C. Zienkiewicz (eds.) – John Wiley & Sons, London 1-20.

Deere,D.U. and Deere,D.W. (1988) : The rock quality designation (RQD) index in practice . *In* Rock Classification Systems for Engg purposes, L. Kirkaldie(ed.), ASTM Spl Publ. 984, 91-101, Philadelphia.

Deere, D.U. and Miller, R.P. (1966) : Engineering classification and index properties for intact rock, Tech. Rep. No. AFWL-TR-65-Air Force Weapons Lab., Air Force Base, New Mexico.

Deere, D.U. and Miller, R.P. (1966): Engineering classification and index properties of rock, Tech. Rep. AFNL-TR-65-116, Air Force Weapons Lab, Alburquerque, NM.

Deere, D.U., Hendron, R.J., Patton, F.D. and Cording, E.I. (1967): Design of surface and near-surface constructions in rock, Proc. 8th U.S. Sym. On Rock Mechanics, New York, 237-302.

Engineering Geology Working Party Report, 1995, Geol. Soc. of London.

Gillott, J.E. (1975): Alkali-aggregates reactions in concrete, Engg. Geology, Vol 9, 303-326.

Grimstad, E. and Barton, N. (1993) : Updating the Q-system for NMT, Proc. Intl. Sym. On Sprayed Concrete- Modern Use of Wet Mix Sprayed Concrete for Underground Support, Norwegian Concrete Association., Oslo, 45-66

Hadley, D.W. (1964) : Alkali reactivity of dolomitic carbonate rocks, Highway Res. Rec., 45, 1-20.

Handin, J. (1966) : Strength and ductility *in* Handbook of Physical Constants: S.P.Clarke (ed.) – Mem. Geol. Soc. America, 97, 223-289.

Hausmann, M.R. (1989) : Modifications at depth by grouting *in* Engineering Priciples of Ground Modification, 346-381 – McGraw Hill Publ. Co.

Heath, R.C. (1998): Basic Groundwater Hydrology, Water Supply Paper 2220, USGS.

Hoek, E. (2000) : Practical Rock Engineering, Rock science Inc., U.S.A.

Hukku, B.M. (1975) : Bhakra Nangal Project, Punjab, Misc Pub. No 29,(Engg. Geol. Case histories), G.S.I.1–10.

Hukku, B.M., Chaturvedi,S.N. and Ashraf,Z (1971) : Tunnelling experiences in Jammu and Kashmir Himalayas – A geotechnical evaluation, Seminar on Geotechnical and Engineering Problems in Tunnelling, Pt.I, New Delhi, 116-127.

Hukku, B.M. and Srivastava, A.K. (1975) : Obra Dam Project , Uttar Pradesh, Misc. Pub. No. 29,(Engg. Geol. Case histories), G.S.I., 11-24.

Iliev, I.G. (1967) : An attempt to estimate the degree of weathering of intrusive rocks from the physico-mechanical properties, Proc. 1st Cong of Intl. Soc. Of Rock Mechanics, Lisbon, 109-114.

Irfan,T.Y. and Dearman, W.R. (1978) : Engineering classification and index properties of weathered granite, Bull. Intl. Assoc. Engg. Geol., No.17, 79-90.

Jai Krishna (1992) : Seismic zoning maps of India, Current Science, Special Issue on Seismology in India,Vol. 62, Nos. 1&2, 17-23.

Jauhari, V.P. (1999) : Operation, monitoring and decommissioning of large dams in India, Draft for Thematic Review, World Commission on Dams.

Kakade, D.G. (1971) : Problems of underground excavation in L. Himalayas with special reference to Yamuna Hydel Project, Stage II, Pt.1, Sem on Geotechnical and Engg Problems in Tunnelling, Pt II, New Delhi,75-92.

King, E.H. (1996): Rock Tunnels *In* Tunnel Engg Handbook., J.O.Bickel, T.R. Kuesel and King,E.H(eds.), Chapman and Hall , 122-152.

Krishnamurthy, J., Manavalan,P. and Saivasan,V. (1992) : Application of digital enhancement techniques for groundwater exploration in a hard-rock terrain, Intl. J. of Remote Sensing, 13(15), 2925-2942.

Kumar A. and Singh K. (2003): Indian stone industry – An insight, Workshop on Marketing and Technology Issues related to Stone Industry, Jan 31.

Kulkarni,M., Nisha ,R. and Deepa ,R. (2004) : GPS and structural monitoring : Deformational studies of Koyna dam, GPS World, Nov.1, 2004

Lumb, P. (1962) : The properties of decomposed granite, Geotechnique, 12, 226-243.

McArthur, J.M., Raven Scroft, P. , Safiullah,S. and Thirlwall, M.F. (2001) : Arsenic in groundwater: Testing pollution mechanisms for sedimentary aquifers in Bangla Desh, Water Resources Research, 37(1), 109-117.

Mahar, J.W., Parker, H.W. and Wuellner, W.W. (1975): Shotcrete practice in underground construction, U.S. Dept, Transportation Rep, FRA-OR&D, Natl. Tech. Inf. Service, Springfield,VA, 75-90.

Mather,B. (1974) : Developments in specification and control, Transportation Res. Rec. , 525, 38-42.

Menon, A.K.M., Gupta, A., Varshney, R. and Uma, B. (1998) : Design feature of power house cavern of Sardar Sarovar Project (Gujarat) – A case study, Intl. Conf. on Hydropower Development in Himalayas, Shimla, 347-356.

Muir-Wood, A.M. (1972) : Tunnels for roadways and motorways, Quart. Jour. Engg. Geol., 5, 111-126.

Nayak, S, Bahuguna, A., Chauhan, P. and Rao, R.S. (1997) : Remote sensing application for coastal environmental management in India, MAEER's PUNE Journal, Spl. Issue on Coastal Environmental Management, 4 (15 & 16),113-125.

Nayak, S. (2003) : Use of satellite data in coastal mapping, Indian Cartographer, CMMC-01, 147-156.

Nelliyat, P., (2000) : Appasamy, P. P., Jayakumar, N. and Manivasagan, R. (2000) : Economic assessment of environmental damage : A case study in industrial pollution in Tiruppur, Environmental Economics Research Committee Project, Madras School of Economics.

Nelliyat, P., (2003) : Industrial growth and water pollution in Noyal River Basin, India, 13th Stockholm Water Symposium, Stockholm.

Obert,L. and Duvall,W.I. (1967) : Rock Mechanics and Design of Structures in Rocks – John Wiley & Sons.

Oliver, H.J. (1979) : A new engineering geology rock durability classification, Engg. Geology, 14,255-279.

Patwardhan,A.M. (1995) : Seismicity impact in Patan Taluks, District Satara, Maharashtra, J. Geol. Soc. India, 46, 275-285.

Peltier, L. (1950) : The geographical cycle in preglacial regions as it is related to climatic geomorphology, Annals of Asso., of American Geographers , 40, 214-236.

Pradeep Kumar and Ajay Kumar Sinha (1999) : Environmental impact assessment of Tehri dam project- A case study, Proc. Natl. Sem. on Environmental Hazards and Sustainable Development, Hamirpur(H.P.) , 51-56.

Rahman, M.G. (1986) : Social environmental effects on large dams, Goldsmith,E. and Hillyard,N.(ed) , Ecosystems Ltd. , Camelford, Cornwal ,2, 267-275.

Rao,D.M. and Gokhale,K.V.G.K. (1973) : Ripple marks in quartzites from Kurnool supergroup Cuddapah Basin, India, J. Sedimentary Petrology , 43(4), 1122-1124.

Rao, G.S.M. (1975) : Rihand Dam Project, Uttar Pradesh, Misc. Pub. No. 29, (Engg. Geol. Case histories), G.S.I.,25-30.

Reddy , C.V.S , Dangwal, M. and Bhattacharya, A. (1998) : Coal mine fire mapping using high resolution airborne scanner data over Jharia Coalfield, Bihar, *in* Remote Sensing in Geosciences, Nitin Kumar Tripathi and Viswa Nath Bajpai(eds), Anmol Publications Pvt. Ltd., New Delhi, 229-236.

Roy Mathew (1996) : Comparative study of hydroelectric projects in Canada and India, Project Report Media Fellowship Programme, Shastri Indo-Canadian Institute.

Sarkar, S.K., Bhattacharya,A. and Bhattacharya,B. (1999) : The main Ganga of North India, an appraisal of its geomorphic and ecological changes, Dept of Marine Sciences, Univ. of Calcutta.

Seshagiri, D.N. (1975) : Idukki hydroelectric project, Kerala, Engg. Geol. Case histories, G.S.I., 238-247.

Simkin,T. and Siebert, L. (1994) : Volcanoes of the World. Geoscience Press, Tuscon , Arizona, 349pp.

Singha, D.R. and Gupta,P.N. (1982) : Assessment of siltation in Tehri reservoir, Proc. International Symp. On Hydrological Aspects of Mountainous Watersheds, Vol.1, Roorkee.

Srinivasan, P.B. (1975) : Koyna hydroelectric project, Stage-I, Maharashtra - Misc. Pub. No. 29, Engg. Geol. Case histories, G.S.I., 31-40 .

Srivastava, M.L. (1974) : Hydrogeological condition in Banda District, U.P, Report of CGWB.

Srivastava, A., Tripathi,N. and Gokhale,K.V.G.K. (1997) : Mapping groundwater salinity using IRS-1B LISS II data and GIS techniques, Intl. J.of Remote Sensing , 18(13), 2853-2862.

Strakhov, N.M. (1967) : Principles of Lithogenesis – Vol 1 (Translated : J.M. Fitzemu), Oliver-Boyd , Edinburgh.

Talwani, P. (2000) : Seismogenic properties of the crust inferred from the recent studies of reservoir-induced seismicity – Application to Koyna, Current Science, Vol.79, No.9 and 10, 1327-1334

Talwani, P (1997) : Seismotectonics of Koyna-Warna area, Pure and Applied Geophysics , 150, 511-550.

Taylor,G.C.(1959) : Groundwater provinces of India, Econ.Geol.,Vol. 54, no.4, 683-697.

Terzaghi, K. (1946) : Rock defects and loads on tunnel support. Rock Tunnelling with Steel Supports, R.V. Proctor and T.White (eds), Commercial Shearing Co., Youngstown, OH, 15-99.

Theis, C.V. (1935): The relation between lowering of the piezometric surface and the rate and duration of the discharge of a well using groundwater storage, Trans. Am Geop Union, Vol. 16, 519-526.

Wickham, G.E., Tiedemann, H.R. and Skinner, E.H. (1972): Support determinations based on geological predictions, Proc. North Amer. Conf. on Rapid Excavation Tunneling, Chicago, 43-64.

Wood, D.F.(1992) : Specification and application of fibre-reinforced shotcrete, Proc. International Symp. On Rock Support, Sudbury, 149-156.

Wright,L.D. (1985) : River deltas *in* Coastal Sedimentary Environments Davis,R.A., Jr(ed), Springer Verlag, N.Y.

Index

A

B

C

D

E

F

M

N

O

P

Q

R

S

T

U

V

W

www.ingramcontent.com/pod-product-compliance
Ingram Content Group UK Ltd.
Pitfield, Milton Keynes, MK11 3LW, UK
UKHW061533190726
13845UKWH00005B/84

9 789352 300655